MADE
ON THE
ISLE OF
WIGHT
FROM TORPEDO BOAT TO SPACECRAFT

BY DAVID L. WILLIAMS

The History Press

Cover illustrations courtesy of British Hovercraft Corporation, Britten-Norman and Westland Aerospace.

First published 2016

The History Press
The Mill, Brimscombe Port
Stroud, Gloucestershire, GL5 2QG
www.thehistorypress.co.uk

Reprinted 2017

British Library Cataloguing in Publication Data.
A catalogue record for this book is available from the British Library.

ISBN 978 0 7509 6754 9

Typesetting and origination by The History Press
Printed and bound in Great Britain

MADE
ON THE
ISLE OF
WIGHT

CONTENTS

ACKNOWLEDGEMENTS

As always, the realisation of a project such as this with its broad scope of coverage can only be achieved with the support, assistance and contributions of many enthusiastic and knowledgeable individuals. I wish to express my warmest appreciation here to those persons who have encouraged me and assisted me with essential information or illustrations:

Tony Baker, Chris Bancroft, Lisa & Phil Beaney, Alan Dinnis, Mike Edwards, John Farthing, John Fisher, Brian Greening, Philip Jewell, John Margetson, Dave Robinson (Aviation Ancestry website), Steve Smith, John Smythers, Graham Stevens, Charlie Taylor, the late Ron Trowell, Steve Tutton, Bob Wealthy, Helen & Colin Wilkinson and Adam Younger. Other contributors, of whom there were many, have been credited by their photographs.

Also, to the following companies, societies and organisations: Air Sea Rescue Marine Craft Section Club (Joe Thomas), Beken of Cowes, Belgian Commando Museum (Richard Schepkens), Britten-Norman (Sheila Dewart), Britten-Norman Aircraft Preservation Society, Classic Boats Museum (Rosemary Joy), Combined Military Services Museum (Nick Turner), Espacenet, Fleet Air Arm Museum (Susan Dearing), GKN Aerospace (the late Tony Roden), Griffon Hoverwork (Selina Kefford), Isle of Wight Heritage Services (Caroline Hampton, Corina Westwood & John Fletcher), Lloyd Images (Mark Lloyd), National Museum of the Royal Navy (Heather Johnson), RNLI Engineering Department, Scottish Maritime Museum, United States Navy and Vestas Sailrockets (Helena Darvelid & Paul Larsen).

As usual, my good friend and co-author on other book projects, Richard de Kerbrech, went to great lengths to assist me, searching out reports and documents to help clarify technical points. I also want to acknowledge with gratitude his valuable explanations of technical terms to improve the accuracy and clarity of the text.

I would like to particularly make mention of John Ackroyd and Raymond Wheeler, two of the high-profile design engineers living here on the Island, both of whom have been immensely supportive to me with information and illustrations of the various projects in which they have been involved. I was a member of Ray Wheeler's staff for some sixteen years of the time while he was the chief designer at British Hovercraft Corporation and later Westland Aerospace.

Finally, I must thank Amy Rigg at The History Press for her endorsement of this project and her support throughout the period while it was being written.

Every effort has been made to correctly attribute the illustrations. If there have been any errors or omissions I sincerely apologise.

INTRODUCTION

For its size and population, the Isle of Wight can claim more records in respect of industrial invention and innovation, and a greater number of record-breaking accolades, than probably any other comparable area in the rest of the United Kingdom, and involving many different product types. But what is more surprising is that, unlike the Midlands, the North East or the Clyde Estuary, and other such industrialised regions, the Isle of Wight is not a place that would be routinely associated with manufacturing industry of any kind.

As if to bear this out, a visitor to the Island was recently overheard on the ferry to say to a companion: 'They have never made much of anything on the Isle of Wight.' What then did he think once emerged from those great Union Jack-painted doors on the large building that is passed as the cross-Solent ferry arrives at East Cowes? And what was the purpose for erecting the great hammerhead crane that is now a familiar landmark in West Cowes? The answers to these questions and much more besides will be revealed in the pages that follow.

So what has been made on the Isle of Wight? The fact is that in the form of shipbuilding, industry was first established on the Island – in the true sense of the word: concerted production on a significant scale and beyond local need – at the beginning of the nineteenth century, prompted by the growth of the navy at Portsmouth and later the emergence of the great port of Southampton. The several existing small yards that were gradually absorbed into a new enterprise had been sporadically building ships of modest size, some for the Royal Navy, for over a century, back to the late 1600s. As a consequence, the company that brought these disconnected facilities together into a single organised concern could later rightly claim to be the oldest shipyard on the Admiralty list, having been in continuous production for 300 years.

This embryonic business, which eventually became known as J. Samuel White & Company, exhibited from the outset the qualities which have become a fundamental dimension of Island industry as a whole ever since that time: conspicuous innovation, ingenuity in design, wide diversity of products through the transfer of technology for other markets and the pioneering of revolutionary production methods. And, along the way, the breaking of numerous records.

White's led the way in the fabrication and testing of models to prove the performance of full-scale craft before they were built so as to ensure the intended functionality and behaviour of their designs. By these means, it achieved significant breakthroughs in composite construction methods, in the development of improvements to marine propulsion systems and in the enhancement of vessel manoeuvrability.

This spirit of enterprise continued into the twentieth century when, following the construction of the first successful flying machines, an emergent aeroplane industry was established on the River Medina, specialising in marine aircraft – flying boats and float planes – exploiting lightweight but immensely strong composite materials that had been developed originally for river craft and, later, high-speed boats. Thus, a sequence of record-breaking hydroplane racing boats originated on the Island with, in parallel, some of the first practical civil and military aircraft. By the 1930s, the industrial mix on the Isle of Wight was a combination of shipbuilding, aircraft production and boatbuilding. The two principal manufacturing companies, White's and Saunders-Roe, were, by then, respected as prestigious manufacturers of international renown. White's had gained an enviable reputation for the destroyers it designed and built, while Saunders-Roe, formerly Saunders, could claim four world water speed record boats, two winners of the coveted British International Harmsworth Trophy, an aircraft that had won the Mortimer Singer prize and others that had achieved record speed and endurance feats. Following a series of experimental prototypes, it was also moving into serial production of small and medium-sized flying boats.

The nature of the Second World War, requiring rapid innovation to enhance the performance of existing military equipment and develop new technologies as the means of defeating the enemy, was the catalyst for an explosion of research and development in local industry. Embracing aerodynamics, electronics, propulsive power and the introduction of modern assembly practices, much of that has continued to the present day. With its already established record in pioneering new concepts, Isle of Wight industry was ideally placed to benefit from these opportunities so that it became the place where many groundbreaking research projects took place. In a sense, in the immediate post-war period, it became something of a design and development hotspot for high-performance jet aircraft, electronics devices and new types of marine vehicle as well as space rockets.

Mass production and long assembly lines are not dimensions of industry that would routinely be associated with the Isle of Wight, the scale of the facilities here precluding such levels of output. Nonetheless, orders for one-offs and small quantities were always problematic in that they generated low returns relative to development and start-up costs and they exposed the workforce to vacillations in demand. The Island's manufacturing businesses, which had benefitted from development contracts, also aspired to derive a greater volume of output of the products they had proven as prototypes both to provide the kinds of remunerative return their investors expected and to secure longer term employment security for their staff. Always vulnerable to the effects of erratic orders, a cycle of recruitment and redundancy often typified the job prospects in the local industrial economy.

In a sense, the determination to generate more sustainable business models has increasingly shaped the Island-based industry of the present day. Thus, ships and hovercraft are no longer built on the Isle of Wight, the construction of complete aircraft is now largely a thing of the past and boatbuilding has declined dramatically despite the convenient access to the sea that is readily available along the shores of the Medina and around Bembridge Harbour. In its place, the Island's industry of today is geared more towards quantity production of two main types of assemblies and components: sophisticated composite structures for a number of sectors, everything from aircraft wing and engine assemblies, smart structures, wind turbine blades and road vehicle bodywork sections fabricated in combinations of exotic materials, and advanced electronics products such as radar systems, navigational aids, modular entertainment packages and a variety of measuring, monitoring and control devices. While the building of new boats has shrunk considerably, there remains a strong and sustained presence in the form of progressive yacht and boat design, largely destined, however, for manufacture by constructors located elsewhere.

In particular, GKN Aerospace in East Cowes continues to have a strong industrial presence, producing a wide range of aircraft engine nacelles and pylons, fuselage, wing and empennage components, winglets and other structures besides electronic components all involving design, test, manufacture, certification and after-market support. Although it is now part of a large group with other manufacturing facilities elsewhere in the UK and world, it has been reliably stated that virtually every large commercial aircraft today uses or has built into it a GKN Aerospace product of some sort – the equivalent of thousands of flights every day with GKN Aerospace technology on board, much of it having originated on the Isle of Wight.

However, the industrial changes that have occurred on the Island have not been achieved without a cost. With the focus on longer runs and repeat orders, much of today's production is lower-tech compared with the past and the levels of pioneering and innovative research have fallen considerably. Also the numbers of employees required along with their skill levels are somewhat lower than they used to be. Inevitably, despite the employment opportunities available from companies like GKN Aerospace, British Aerospace, Britten-Norman, Vestas, Vikoma, Gurit, PEC and others, the situation today is a lot less positive than it once was.

The industrial component of the Island's economy has been steadily shrinking for some time, part of a continuing decline that began in the mid-1960s with the closure of the J. Samuel White boatbuilding and shipyard businesses with the loss of many valuable skilled jobs. The size of the industrial workforce of the early 1960s, itself a marked reduction on the numbers that were employed back in the late 1940s, had more or less halved in fifty years, a severe reduction. Of course, it is not just the Isle of Wight that has been affected by this. Regrettably, it is part of an accelerating national trend arising from the adoption of economic strategies that have meant the abandonment of Great Britain's long held status as 'workshop of the world'. The rather short-sighted strategic decision to effectively relinquish manufacturing industry to concentrate on financial services has had a detrimental impact on the UK as a whole.

Here on the Island the impact of such policies has also witnessed the near eradication of quality indentured apprenticeship schemes (not to be confused with many of the so-called apprenticeships of today) for youngsters now entering the employment market. Today youth unemployment on the Isle of Wight remains stubbornly high, amongst the worst in the country; a problem which is not being helped by insufficient good engineering and trade apprenticeships. In the past an apprenticeship, such as the superb schemes once offered by White's and Saunders-Roe, meant a truly worthwhile start to working life for a young person, a thorough indentured course of trade training from two to five years' duration under the guidance of skilled craftsmen leading to a respected and recognised qualification.

The absence of adequate skilled and appropriately remunerated jobs in general has also resulted in a low-wage economy on the Island which in turn frustrates advancement and depresses infrastructure improvement and investment in the local community. It is disappointing to think this should be the situation at the time of writing given the Island's incredible industrial heritage and pedigree.

Past experience has also demonstrated the vulnerability of the Island's industries to the whims of politicians of all persuasions. Institutional investors are often reluctant to back cutting-edge technology so government support was invariably essential to help get new products established. It is a question of the national interest, helping the United Kingdom hold its own in an increasingly competitive world. Some inspired technological achievements with potentially rewarding business prospects for the country as a whole have been lost through failures to provide committed and enduring support and sustained investment, all too often reflecting a failure of political will. Notable among the victims of such fickle backing here on the Isle of Wight were the SR.177 aircraft programme, killed by a misguided defence review; the Black Arrow satellite launcher, cancelled even before its greatest success, leaving Britain entirely dependent on foreign space vehicles; and the once promising hovercraft industry, beset by crippling bureaucracy (was it a ship or a plane?) and foreign policy vacillations. Britten-Norman, too, with its world-beating Islander aircraft struggled financially at times because of the unwillingness of city financiers to inject capital into the business.

Despite these setbacks, they do not detract from the Isle of Wight's immense industrial and manufacturing record or the quality and worth of its many achievements. The Isle of Wight has exhibited, and still does exhibit, extraordinary innovation in design and engineering, an attribute that goes back over almost two centuries. *Made on the Isle of Wight* traces industry on the Island from pre-Victorian times to the present day, highlighting the numerous amazing, sometimes groundbreaking products whose origins were often in grimy workshops, boat sheds and small factories on this small island.

There is so much to present here in this book that concentrates on the Isle of Wight's vehicle and transport-related products. But inevitably, in such a limited space, some examples of the Island's diverse output have had to be omitted although it is hoped that nothing significant has been missed.

The diversity of the projects pursued by Saunders-Roe alone in the period from 1945 to 1965 would fill volumes. For interest, and because the story would not be complete without demonstrating the extent of the forward thinking exhibited by the various designers and advanced projects teams, some examples of their unfulfilled schemes have also been included. These were not pipe dreams as might be thought but serious design concepts, devised for specific purposes and subjected to examination of their technical practicality and, in most cases, their commercial viability too. Certainly, some were intended solely as feasibility studies requested by potential customers. Some were expressions of intent with regard to the replacement of existing products with improved or enlarged versions only for them to be undermined by unfavourable commercial or political conditions. Others failed to materialise simply because the funding dried up as budgets were cut. Indeed, in more than one instance, craft in the process of construction as part of high-profile programmes were cancelled and abandoned incomplete.

Not only does *Made on the Isle of Wight* present the many diverse and novel creations, inventions, innovative concepts and record-breaking achievements that the Island was, and is, proudly responsible for but it also briefly relates the stories of certain of the designers and engineers who conceived them and the businessmen who set up companies to make them. Likewise, the teams of loyal and committed work people who fabricated the physical end products are celebrated here visually in a dedicated chapter.

It is hoped that this book will be a fitting tribute to the ingenuity that has blossomed on this Island and to the engineers, designers and many workers who were and still are employed in local industry. It is hoped, too, that it will be appreciated as a revealing showcase of the Isle of Wight's incredible industrial heritage by both the casual reader and everyone who is interested in engineering and design history.

Note: Although the town of West Cowes was officially renamed Cowes in August 1895, for clarity, the old name of West Cowes has been used throughout this book wherever there has been a need to make a distinction from the twin town of East Cowes.

I

INGENUITY & QUALITY
The Dawning of an Age

George III was on the throne, Nelson's triumph at the Battle of Trafalgar was two years in the future and it was barely twenty-five years since James Watt had produced his first practical steam engine, launching the Industrial Revolution. The year was 1802, around the time when the White family began to transfer its boatbuilding business from Broadstairs, Kent, to Cowes, Isle of Wight, and establish the shipyard there that would take Island shipbuilding into the modern age. By taking over the small existing yards, expanding them and investing in major new facilities, including dry docks and additional slipways, the shipyard's capacity was both extended and enhanced. Later, a process of rationalisation radically improved the construction workflow, permitting the company to derive the greatest benefit from the resources at its disposal.

While this programme of modernisation was gathering pace, efforts proceeded in parallel to introduce improvements to ships, an important dimension of White's business philosophy which was to continue into the future, throughout the next century and a half. To that end, John White, the son of Thomas White who had instigated the move to the Island, had a model-testing pond created in the garden of his father's house in which clockwork models of new vessel designs were evaluated before the commitment was made to full-scale construction.

As a matter of interest, apart from John White's testing pond, no fewer than five other test tanks were later constructed on the Isle of Wight for the evaluation of models of marine craft. The John I. Thornycroft tank at Steyn Battery, Bembridge, measuring 60ft in length was privately opened in 1910. Four more of varying lengths were commissioned by Saunders-Roe (later British Hovercraft Corporation) at its Osborne Works research and experimental site at Whippingham, among them the 618ft long No. 1 tank opened in 1947 and the even longer 650ft No. 3 tank built in 1969, described as a world-class facility. Closure of the latter in 2008 was regarded as a serious blow to the UK's fluid dynamic's industry. The other tanks have also been decommissioned, a tragic loss of superb research facilities.

A number of key improvements to hull design and construction methods emanated from John White's pioneering experiments. Among them was a vessel widely credited as having been the first to exhibit clipper lines, a patented system of double-diagonal hull assembly initiated in response to a request from the infant P&O Line, and a non-capsizeable lifeboat, discussed in more detail later, which was produced in several variants in greater numbers than probably any other comparable craft.

As the nineteenth century unfolded, numerous other patents were registered by the company, demonstrating a great diversity of inventiveness and ingenuity. Some of these passed without implementation, others were overtaken by events as the pace of industrial progress accelerated but the quantity alone revealed the inventive enterprise which characterised White's business. The shipyard later became a world leader in the building of naval destroyers, a distinction which in part originated through its introduction of the patented turnabout steering system for torpedo boats, an artifice which made these craft extraordinarily manoeuvrable. White's next conceived the torpedo boat catcher, the logical counter weapon to those menacing craft and a warship type that evolved first into the torpedo boat destroyer and then, ultimately, to modern destroyers in all their forms. From small escorts through to larger flotilla leaders, White's at Cowes built them all.

Towards the end of the Second World War, arising from its respected and prestigious reputation as an Admiralty builder of top-quality vessels, the shipyard was selected for the inception of the full welded construction of destroyers. Resulting from this, HMS *Contest* became the Royal Navy's first all-welded destroyer.

Besides these advances in ship design and assembly, White's remained active throughout its existence looking for other applications for its technological concepts, adapting products for other related markets. These featured prominently in the output of the Engineering Department, dealt with in Chapter Three, but also in other marine products such as flying boat pontoons, water scooters and hydrofoils, pilot hoists and composite indicator buoys to highlight just a few. And it wasn't only products for maritime purposes. When times were tough as in the Depression, also following the two world wars and in the early 1960s, when British shipbuilding went into steep decline, the company turned its hands to almost anything, contriving such implements as air-conditioning units and even ice-cream makers and milk coolers.

The purpose of this book is not only to showcase the many innovative, groundbreaking products of Isle of Wight industry but also to convey something of the personal backgrounds and accomplishments of some of the leading designers and engineers whose inventive skills brought them into being. For White's there were a number of such people but here the focus is on the leading light of the White family itself. Thomas White encouraged all of his sons, Joseph, Robert and John, to take an active role in the family business which they did with evident enthusiasm and commitment – the origins of the boats completed between 1820 and 1860 revealing their individual and shared enterprises and specialisations. Through the active involvement of that generation, the White business was able to move forward on a broad front and expand in a way that might not otherwise have been possible. Yet it was from the next generation that all that energy and progress was harnessed into a single organisation that could rank with the best across the country and in the wider world.

It was John Samuel White, son of John and grandson of Thomas, who may be regarded as the father of the company which carried his name alone for eighty years from around 1885. Born in 1838, he entered the business as a young man and the records show that already by the age of 20 he had been entrusted with the running of the Falcon Yard and the lifeboat shops in East Cowes. Subsequently, he consolidated all the shipbuilding activities on both sides of the Medina under single management. He launched the company's official yard list and oversaw all the major developments as the business became, besides a shipbuilder, a marine engineering concern and an aircraft manufacturer.

He was the inspiration for the rationalisation of the yard's building facilities, having hull construction concentrated on slipways in East Cowes supported by plate-cutting, casting, frame bending and other workshops, with fitting-out, machining and engine fabrication as the primary functions on the west shore of the River Medina, again arranged with all the facilities necessary for the required trades. Through acquisitions and the adherence to and widespread promotion of the highest standards of quality, even at the risk of losing contracts when it was felt that such principles would be compromised, he turned White's shipbuilders into an internationally recognised and respected business.

His sons, John Lee White and Herbert Samuel White, entered the business in 1884 but after John Samuel's death in 1915, the company, which by then was a publicly listed firm (it had been constituted as a limited liability company on 11 February 1898), passed out of the White family's hands.

From around 1865 to 1965, White's produced more than 2,500 ships, boats and other marine craft. By 1919, over 4,500 men and women were employed in the shipyard, a level of employment that remained more or less constant through to the late 1940s. In the early 1960s, by which time employment numbers had virtually halved, White's was caught up in the general slump in world shipbuilding. The effects of a sharp reduction in new orders was exacerbated by what had been a protracted lack of investment in modern prefabrication facilities since 1945, although it has to be said that the company was to some extent constrained geographically, limiting the scope for major redevelopment. This led to the closure of the boatbuilding operation in 1964 and the entire shipbuilding business a year later. It was at the very time when the demand for off-shore supply craft (OSSV) for North Sea oil and gas exploration was about to soar; these ships well within the yard's capabilities an example of which, the *South Shore*, had in fact been delivered in that very same year. The Engineering Department was all that survived, continuing to operate for another sixteen years, but that is a story for a later chapter.

Launched at East Cowes on 18 June 1932, the 90ft brigantine *Waterwitch* was the third of three sailing vessels designed and built by Joseph White for the Earl of Belfast, the others being the *Therese* and *Louisa*. It is claimed that the *Waterwitch* was the first vessel to exhibit the classic hull lines that later became famous in the great trading clippers, as evinced in her long, low hull, her hollowed bow and sharp-raked stem and her overhanging counter stern, features based on the design concepts of Thomas Assheton-Smith. A contemporary description stated that her 'innovative design did much to advance naval ship performance'. After defeating all the opposition in competitive races, she was acquired by the Admiralty and employed pursuing slave traders off the West African coast until disposed of in 1861. (Victor Collypriest, courtesy J. Samuel White)

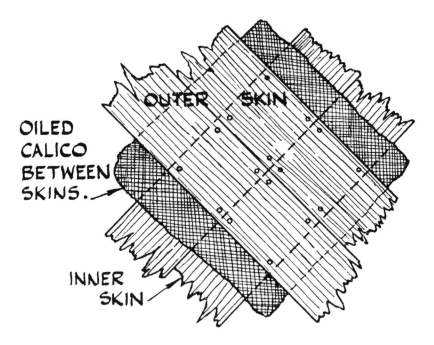

White's introduced a groundbreaking construction concept in the mid-nineteenth century, a patented double-diagonal planking system of composite construction – double-skinned timber strakes with a waterproof membrane in between, set at opposing angles over wrought-iron frames. It earned the company prestigious orders for packet steamers for the infant P&O and Royal Mail Lines. (Richard de Kerbrech)

A profile view of RMS *Solent*, the largest ship ever built by the double-diagonal wooden system when completed in 1852. Although iron hull construction rapidly overtook double-diagonal wooden construction for large ships, it remained in continued use for small craft well into the next century. At 2,230 tons and 310ft in length, Royal Mail's *Solent* was also the largest ship built on the River Medina up to that time. (Nigel Robinson)

The White's built HMS *Swift*, completed in 1885, may be regarded as the ship from which future naval destroyer development largely originated. Her design recognised the need for a vessel to counter torpedo boats. Thus, she was conceived as a torpedo boat 'catcher' and as such she represents a key milestone in the development of the first true torpedo boat destroyers. Built as a private venture, incorporating White's patented 'Turnabout' steering installation of rudders fore and aft of the propeller, she displaced 137 tons and had an overall length of 154ft. Steam-reciprocating engines rated at 1,330ihp gave her a top speed of 23.75 knots (27.3mph). Purchased by the Royal Navy within a year of completion and designated TB81, the *Swift* remained in commission until sold for breaking up in October 1921. (Crown Copyright)

Above: A close contemporary of HMS *Swift* but somewhat larger was the 350-ton, 210ft-long torpedo boat catcher *Sea Serpent* laid down in 1887, another of White's private venture vessels. After purchase by China in 1894, and when she was renamed *Fei Ting*, she was modified and strengthened by Armstrongs at Elswick. The photograph shows an unidentified White's torpedo boat off Cowes thought to be the *Sea Serpent* prior to being rebuilt. (Isle of Wight Heritage Services)

Opposite top: White's revolutionary Turnabout principle, patented in 1882 and first incorporated in a launch of that name built for the Calcutta Port Trust, gave vessels extraordinary manoeuvrability; they could virtually turn around in their own length. Its success led to a glut of Royal Navy orders – a total of forty-seven torpedo boats and torpedo boat destroyers by 1914. Shown here in the entrance to Cowes harbour is Turnabout Torpedo Boat 34, delivered in August 1886, one of five such craft. Lighter and shorter than HMS *Swift* at 125ft, TB34 retains the twin side-by-side arrangement of her funnels. (Debenhams, courtesy of Isle of Wight Heritage Services)

Above: Whereas many Admiralty orders were for standard-class destroyers, White's also received orders from foreign navies both before and after the First World War to build ships of this type to its own designs. Such vessels were often superior to their Royal Navy counterparts. One such contract received in 1912 was for six large super-destroyers for the Chilean Navy. Only two were delivered prior to the outbreak of hostilities, the remainder completed for the Royal Navy with modifications. The world's largest destroyers at the time, they displaced 1,695 tons and measured 320ft overall length. The photograph shows HMS *Botha* (ex-*Almirante Williams Robelledo*), third of the four ships operated by the Royal Navy. (Crown Copyright)

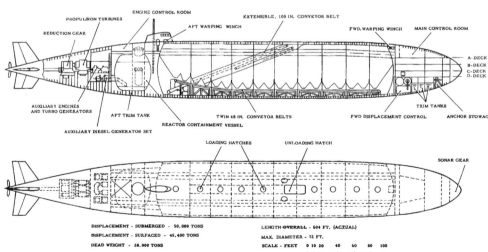

General arrangement of a proposed 50,000 ton nuclear submarine cargo vessel for Mitchell Engineering Ltd. 1959.

Opposite top: Two submarines were also built by White's in the First World War, but a third was cancelled while under construction. This is the second, the F2, completed in July 1917. It displaced 353 tons and was 150ft long. (Crown Copyright)

Opposite bottom: Although no other submarines were built on the Island, work on large undersea craft resurfaced some forty-five or so years later when Saunders-Roe was invited by Mitchell Engineering Limited to carry out a detailed design study for a 50,000 tons displacement nuclear-powered submarine cargo vessel. Designated project P.212, the 604ft-long submarine, with a maximum diameter of 74ft, would have had a cargo capacity of 28,000 deadweight tons. A 6.5ft scale model was evaluated in the company's test tanks and the research indicated that the concept was viable. However, although a detailed design order was contemplated in conjunction with a major shipyard, nothing materialised. (Saunders-Roe, courtesy of Ray Wheeler)

Below: Between the wars, White's was called on again to design and build exceptional destroyers for foreign governments. Among them was the Mendoza class of three flotilla leaders ordered by Argentina. With a displacement tonnage of 1,570 and an overall length of 321ft, they were fast ships and considerably superior to other contemporary destroyers. This is the *Mendoza*; her sisters were the *La Rioja* and *Tucuman*. (Isle of Wight Heritage Services)

Above: White's completed its largest ship ever, the fast minelayer HMS *Abdiel*, of 4,000 tons and 418ft overall length, in 1940. Launching a ship of this size, by necessity at the highest level of tide, required the swift intervention of tugs and stretched the yard's capacity, both on the slipway and the safe distance across the river, to the limit. With a top speed in excess of 40 knots (46.0mph), *Abdiel* and her sisters were the fastest ever Royal Navy surface warships. She was lost on 10 September 1943 when she struck a mine in the harbour at Taranto, Italy. (World Ship Society)

Opposite top: With another war imminent, warship orders for both Royal Navy and foreign navies increased. Among them were the two most outstanding vessels of this type completed by the shipyard, the super-destroyers ORP *Grom* and *Blyskawica* of 1937 ordered for the Polish Navy. Designed by the company, they were more advanced than the equivalent ships of virtually every other navy, including the Royal Navy. They had a standard displacement of 2,140 tons and an overall length of 377ft. As such, they were the longest ships so far launched by the company. This is the *Grom* in Gdynia Roads prior to the war. (Marek Twardowski)

Opposite bottom: The sole survivor of the two exceptional Polish destroyers built by White's, now preserved afloat in Gdynia, is the ORP *Blyskawica*. Her sister ship was sunk on 4 May 1940, when a bomb from a German aircraft detonated a torpedo that had been loaded into her midships launcher. (Marek Twardowski)

HMS *Cavalier*, a part-welded Emergency Ca-class destroyer, is now preserved in dry dock at Chatham, the sole surviving Royal Navy destroyer of the Second World War period. The fore and aft hull sections were welded while her midsection was riveted. (Graham Stevens)

As a tribute to the high quality of its workmanship and experience with two part-welded Ca-class ships, White's was selected to pioneer full welded construction of naval ships in the form of the HMS *Contest* commissioned in 1945, a near sister-ship to HMS *Cavalier*. She was 2,560 displacement tons and 363ft in length. Her construction required the erection of a new welding shop to handle the larger sub-assemblies. The practices developed by White's were transferred to and adopted by other shipyards. (Maritime Photo Library)

Work in progress in White's welding fabrication shop in East Cowes. The photograph shows large aft hull bottom sections under construction for the destroyer HMS *Dainty*. (White's Archives, Isle of Wight Heritage Services)

Not the last ship on the J. Samuel White yard list but the last ship to be completed and delivered by White's prior to the closure of the shipyard in November 1965 was the frigate HMS *Arethusa*. Around 2,580 vessels of all sizes had been constructed by the company since the inception of the official yard list 100 years earlier. *Arethusa* was launched on 5 November 1963, leaving behind empty slipways that would never again be used. (Ron Trowell)

Though not an exceptional ship, HMS *Arethusa*, a Leander-class frigate, was the last in a long line of vessels 'well built' for the Admiralty by White's and its forebears, going back for 300 years. She had a displacement of 2,350 tons and was 373ft long. (Ron Trowell)

An interesting application of J. Samuel White's ship and boatbuilding skills was the construction of buoyant flying boat docking pontoons which were used by BOAC at its Hythe maintenance centre and at the passenger terminus in Southampton's Old Docks. (Keith MacDonald)

2

FROM STRINGBAGS TO SHIPS IN THE SKY

As had been the case with the White's shipyard, so too the next large manufacturing concern to emerge on the Island was also established by an 'overner' (a person not born on the Isle of Wight), a man named Samuel Saunders. Born in 1857, at a young age Sam Saunders took over the running of the boatbuilding firm at Goring on the River Thames in Oxfordshire, which had been opened originally in 1830 in nearby Streatley by his grandfather Moses Saunders. Recognising that in order to expand, the firm needed access to other markets besides small river craft, he partially moved the business to Cowes, Isle of Wight, in 1901, where it could benefit from the growing popularity of yacht racing and leisure sailing. Of importance, he took with him two valuable colleagues from the boatyard team – two of his greatest designers, Sidney 'Joe' Porter and Fred Goatley, of whom more later.

After briefly establishing himself under the name Saunders Patent Launch Building Syndicate in Alexandra Hall in Birmingham Road, West Cowes, in 1910 he took over the Columbine Yard in East Cowes, previously occupied and recently vacated by the Liquid Fuel Engineering Company (LIFU), along with the adjoining Sunbeam Yard, Seaholme building, Esplanade building and Medina shop.

Sam Saunders had developed a method of composite construction of boat hulls, inspired by the birch bark canoes of Native American Indians. The material that resulted from the stitching together of ply veneers with waterproof calico inter-layers had great strength but was light in weight. It was patented in 1898 (No. 222) under the name Consuta and again in modified form in 1906 (No. 21,030).

Utilising Consuta, Saunders' nascent Island concern proceeded to construct a number of fast racing craft, designed as hydroplanes with stepped underwater hulls. So successful were these craft that Saunders earned a reputation for his racing boats and a sequence of ever faster record-breaking race winners subsequently emerged from his boat shop. Among them were two world and European speed record holders, two winners of the prestigious British International (Harmsworth) Trophy and numerous other contenders.

It became apparent that the Consuta material was equally suitable for other purposes, both for other forms of transport as well as for building structures. As a consequence, early on in the fast-emerging age of aviation, the Saunders business completed the hull of a small flying boat for the pioneer aviator Thomas Sopwith for him to enter in the Mortimer Singer prize race of 1911. Named *Bat Boat*, the first of a series, it went on to successfully complete a course over the Solent in July 1913 within the stipulated 5 hours, in which an amphibious aircraft was required to land and take off six times alternately from points at land and sea which were a minimum of 5 miles apart. This venture was the catalyst for the Saunders firm to turn its attention increasingly to flying machines and soon it was developing its own new designs for maritime aircraft.

In parallel with Saunders' progressive involvement with aircraft, the J. Samuel White shipyard also opened an Aviation Department which, unlike Saunders, focused on float plane development rather than flying boats. Under the leadership of Howard Theophilus Wright, a series of aircraft were developed bearing the trade name 'Wight'.

Born in Dudley, Staffordshire, in 1867, Howard Wright was educated at Manor House School in Ashby-de-la-Zouch before entering an apprenticeship in his father's company. After the American-born mechanical and electrical engineer Sir Hiram Maxim acquired part of Joseph Wright's business in 1899 to form the Maxim Electrical & Engineering Company, Howard Wright worked for him for the next five years as works manager. During this time he participated in the development of the American's pioneering though unsuccessful aircraft schemes. Later, from 1909, having set up his own business, he began designing aircraft under his own name and in that year, assisted by designer William Oke Manning, he completed the Howard Wright Biplane, which was distinctive for the time in having framework partly constructed from steel tubing. In 1912, after the Coventry Ordnance Works secured a controlling interest in his concern, he was appointed as the head of the newly formed Aviation Department of J. Samuel White at Cowes and he remained in that position until the closure of the Department in 1919. With the exception of the Admiralty-designed aircraft built by White's and the Short 184 seaplanes constructed under licence, Howard Wright was responsible for the entire range of Wight aircraft produced in those seven years: the Wight Seaplane, Navyplane, Twin, Baby, Landplane Bomber, Quadruplane and Wight Converted Seaplane.

By the time of the outbreak of the First World War, White's had the first of these aircraft designs in production. To some extent, however, the war frustrated the plans of both Saunders and White's and for the duration both companies were engaged by the War Office building machines to the designs of competing aircraft manufacturers. In total Saunders' wartime output amounted to 399 aircraft while White's constructed another 219 machines.

For White's, the war demonstrated that its strength really lay in shipbuilding and, as a consequence, its Aviation Department was closed and its aerodrome at Somerton, on the outskirts of Cowes, was sold off. In contrast, Saunders' future was very much in aircraft fabrication, particularly after an inspirational young designer, Henry Knowler (later called 'Mister Flying Boat'), joined the company.

Having worked previously for Vickers, at the age of 29, Henry Knowler was engaged by S. Saunders Limited in 1923 when the company reinstated an aircraft manufacturing operation, serving initially as assistant to chief designer Bernard Thomson. With the departure of Thomson in 1926, Henry Knowler was appointed chief designer and as such he had responsibility for all flying boat projects undertaken by the company from that time: the Severn, Cloud, Cutty Sark, Windhover, London, A33, Lerwick, Shrimp, SR.A/1 and the Princess. A committed advocate of flying boats for air travel, he was promoted to technical director from 1952, a position he held until 1956 when he retired.

Aircraft and boat production lines were soon functioning in parallel at Saunders, the former concentrated in the new Solent Works, opened in 1916 on the West Cowes shore adjacent to Medina Road, the latter in the Columbine Yard and at the Cornubia Works further up river on the eastern shore, while a dedicated Consuta manufacturing plant, the Folly Works, was established at Whippingham.

Through a series of models – the Kittiwake, Medina and Valkyrie, constructed from wood, and the Severn, London, Lerwick and smaller amphibians, constructed from Alclad aluminium – the company's designs eventually culminated in the giant Shetland, a collaborative project with

Short Brothers of Rochester, and the even bigger Princess flying boats, the latter designed and built exclusively by Saunders-Roe.

During the period of continuous development in the 1920s, the Saunders concern ran into financial difficulties, a setback from which it only recovered after Alliott Verdon-Roe, the man who had created the Avro aircraft company, stepped in and took it over on 10 December 1928 in a partnershop with John Lord, with further investment from the Aircraft Investment Corporation. Thus from 3 July 1929, the East Cowes business was renamed Saunders-Roe.

Simultaneously, the original Columbine Yard was demolished and a new purpose-built Columbine Works was erected as part of a major enhancement of the production facilities in readiness for the commencement of work on the London flying boats ordered by the RAF. Alongside it, completed in 1939, was a new Medina Shop, another large building in which Lerwick aircraft were assembled. By the mid-1930s Saunders-Roe had no less than four manufacturing sites in full operation on the River Medina: Columbine, Solent, Cornubia and Folly with the Somerton aerodrome also by that time in its possession.

Although land planes swiftly overtook marine aircraft in the years following the Second World War, having benefitted during the conflict from accelerated development in the form of long-range bombers and transports, Saunders-Roe continued with ambitious flying boat designs well into the 1950s. The company remained both positive about the continuing viability of flying boats and optimistic about getting the three propeller-driven Princess flying boats into commercial operation. Many schemes for the three giants were pursued some of which involved them being adapted for other purposes. Project P.213, an investigation contracted by the US Navy, even explored the possibility of fitting them with nuclear reactors to drive a combination of power and compressor turbines in order to provide long endurance for oceanic patrol.

But when the flying boat era did finally come to an end, the innovative Saunders-Roe company embarked next upon the construction of jet aircraft, space rockets, helicopters and hovercraft, making a name for itself yet again with all these diverse and futuristic vehicle types.

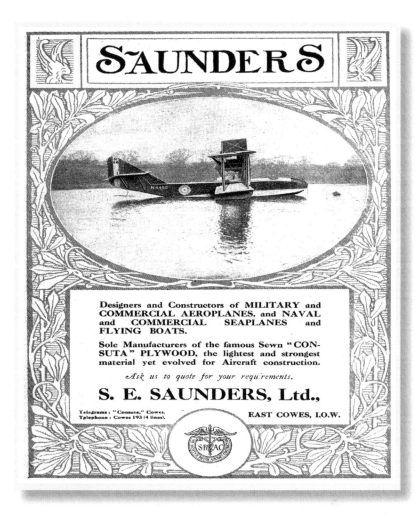

Saunders' rise to prominence as a boatbuilder and later as a constructor of amphibious aircraft owed much to the development of its patented 'Consuta'-sewn plywood. Comprising a wire-stitched sandwich of laminated veneers and calico inter-layers, it is described in this August 1919 advert as 'the lightest and strongest material yet evolved for aircraft construction'. Besides aircraft products, 'Consuta' was used for other applications such as motor-car bodywork, interior panels and prefabricated buildings. The aircraft depicted here is a Felixstowe F2A, 100 of which were built by Saunders in its Solent Works during the First World War. *Aviation Ancestry*

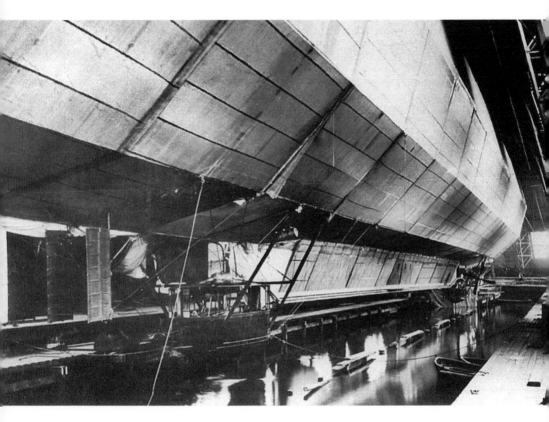

Above: Among the many applications for Consuta was the construction of airship gondolas. In this example they were supplied in 1910 for the first British airship, the HMA1 *Mayfly*, built by Vickers. The fore and aft gondolas can be seen under the airship while it was secured inside its hangar. They were 25ft long and housed the engines and accommodated the crew. (Author's collection)

Opposite top: Subsequently, in 1916, Saunders built the engine gondolas for the airships R31 and R32 for Short Brothers, four in each case. A single 275hp Rolls-Royce Eagle engine was mounted in each gondola. The four gondolas of the R31 can be clearly seen beneath the massive hull of the 618.5ft-long airship as she floats low above the ground at Cardington, Bedfordshire. (Crown Copyright)

Opposite bottom: Saunders' venture into aircraft production began with the hull of the first Sopwith Bat Boat, completed in 1912. Powered by a six-cylinder Green engine producing 100hp and piloted by Harold G. Hawker, this small flying boat went on to win the Mortimer Singer Trophy on 8 July 1913. The hull was fabricated at Cowes from Consuta plywood, the wings and remaining structure assembled at Kingston-on-Thames. Later Bat Boat hulls were not made using Consuta but they were built to Saunders' single-step design. (Saunders-Roe)

Below: Saunders' first fully home-grown aircraft, the flying boat Kittiwake (G-EAUD), shown here in displacement mode under tow, was intended for civil operation carrying six to seven passengers. Responsible for the overall design of the Kittiwake was Francis Percy Hyde Beadle, ably assisted by Sidney (Joe) Porter. The aircraft, which had a two-step hull, was intended as an entrant into an Air Ministry competition for landplanes and amphibians held at Felixstowe in the late summer of 1920. (Author's collection)

Above: The Wight No. 2 Navyplane, designed by Howard Wright and completed in 1913, was a rebuild of the earlier No. 2 Seaplane which in turn was a rebuild of the No. 1 Seaplane, White's first aircraft type. It was a pusher configuration biplane intended for naval patrol, 31.5ft long with a maximum wingspan of 44ft. The sole aircraft of the type is seen manoeuvring in Cowes harbour prior to take off with test pilot Eric Gordon England at the controls. The design was eventually developed into the larger Improved Navyplane of which a total of eighteen were built from 1914. (Debenham, courtesy of Isle of Wight Heritage Services)

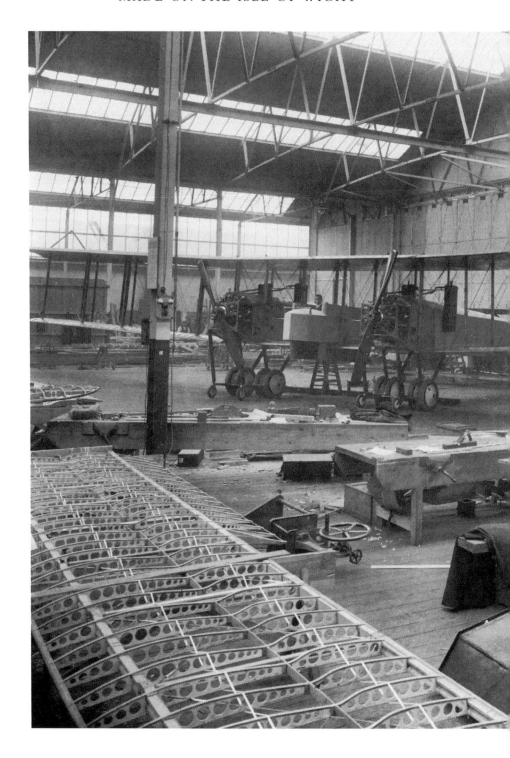

The aircraft under construction in this inside view of the White's Aircraft Department in West Cowes in 1915 is the first of the Wight Twin biplane bombers. Unlike the succeeding three aircraft, which were fitted with floats for service with the Royal Naval Air Service, this machine was completed for the French forces as a landplane with wheeled undercarriage. These aircraft were said to have a range of 400 miles. (Isle of Wight Heritage Services)

Above: The most impressive of the improved marine aircraft types to follow the Navyplanes was the Wight AD Type 840 Seaplane, which was taken up by the Royal Navy with an order for thirty scout aircraft. A follow-on order for fifty aircraft had to be outsourced because White's did not have the factory capacity. The prototype machine, number 835, is seen afloat in the River Medina looking towards the slipways of the East Cowes shipyard. (Crown Copyright)

Opposite top: White's was responsible for the construction of one of the largest aircraft built in the First World War, the twin fuselage AD Type 1000 King Cormorant torpedo bomber, the largest British aeroplane until 1918. Although of moderate size by later standards, it was 64ft long with a wingspan of 115ft and 12.7-ton maximum weight. Alongside it is the much smaller Wight Trainer Seaplane. Of the seven King Cormorants ordered from White's, only two were completed. (Crown Copyright)

Opposite bottom: White's also developed a number of other landplane aircraft, among them the unusual four-winged Quadruplane, seen near the Somerton Works which J. Samuel White's opened in April 1917 specifically for aircraft production. The company even constructed a track-way running down the hill to get finished floatplanes to the River Medina. A replica of the Wight Quadruplane is displayed at the Solent Sky Museum, Southampton. (Crown Copyright)

Above: Although, like Saunders in East Cowes, White's mainly built other company's aircraft during the First World War, it did also proceed with its own designs, the most successful of which, adapted from a Landplane Bomber, was the Wight Converted Seaplane of 1917. Forty of these tractor-propelled floatplanes were delivered to the Royal Navy, one of which was the first aircraft ever to destroy a submarine when, on 22 September 1917, a machine flown by Sub Lieutenant Charles Stanley Mossop, with Air Mechanic Ingledew, bombed and sunk the German UB32. The aircraft shown here is the prototype Converted Seaplane, No. 9846. (Isle of Wight Heritage Services)

Opposite top: Despite the imminent closure of its Aircraft Department, following the departure of Howard T. Wright as chief designer, White's was still advertising itself as a 'Warship and Aeroplane Constructor' as late as 3 October 1918, when this advert appeared in the trade press. (Aviation Ancestry)

Opposite bottom: Working with engineers at the Marine Aircraft Experimental Establishment, RNAS Felixstowe, Suffolk, Saunders embarked on an experimental project in 1921 based on the Felixstowe F5 aircraft. A specially constructed tunnel or channelled hull was made in Consuta to Saunders patented design (No. 230196), conceived by chief designer Bernard Thomson. Essentially a central, inverted 'v' groove running the length of the underside of the hull, this reduced spray at take-off but it carried a weight penalty and the greater area of wetted surface increased drag. (Author's collection)

THE "WIGHT" SEAPLANE
CONSTRUCTED BY

Flying boat construction at Saunders accelerated under the guidance of Henry Knowler (Mr Flying Boat) who joined the company in 1923. Later as chief designer, replacing Bernard Thomson, he took over responsibility for the development of the three-engined A.3 Valkyrie, designed in response to Ministry of Aircraft Production specifications R14/24 and R22/24. Fabricated from Consuta, it was the last and largest wooden machine to be built by Saunders. The photograph shows the almost complete aircraft in the Solent Works in 1927. (Author's collection)

Saunders turned to lightweight ribbed Alclad aluminium for the flying boats built subsequent to the Valkyrie, first producing a one-off experimental Alclad hull for a Supermarine Southampton flying boat, designated the A.14, in collaboration with the Southampton-based aircraft manufacturer. The aim was to produce a strong but lightweight and roomy hull through the use of flat-sided skins with horizontal external corrugations in place of internal stringers. Another Knowler innovation was to give the hull a single V-form curvature, a concept adopted in later aircraft. (Author's collection)

Above: Saunders next produced the A.7 *Severn*, completed in 1927 and seen here on its landing approach near Cowes harbour. Powered by three Bristol Jupiter engines, producing 1,455bhp, the Severn had a maximum speed of 125mph, an increase of 23mph over the comparable but more powerful, wooden-hulled Valkyrie. The Severn was the first RAF machine to fly non-stop from Gibraltar to Plymouth. (Saunders-Roe)

Below: Another unusual air vehicle, which demonstrates the innovative character of Island industry, the Isacco Helicogyre No. 3 as it was called, was built in 1929 by Saunders. It had engines with tractor propellers fitted at the tip of each rotor. Though not of Saunders design, being the conception of Vittorio Isacco for research purposes, and despite its impracticality, the construction of the Helicogyre showed how the Saunders company was prepared to support innovation in order to aid progress in the development of aircraft. (Author's collection)

Above: Stripped of its upper aerofoil and undercarriage and with extra fuel tankage installed, Windhover G-ABJP, shown here, set an assisted endurance record in August 1932 piloted by Mrs Victor Bruce. Twice refuelled in flight to extend its range, a British record of 54 hours and 13 minutes was established. But for the weather conditions, it would have continued for longer. (Author's collection)

Opposite: Saunders-Roe's three small amphibians, the twin-engined A.17 Cutty Sark and A.19 Cloud, and the three-engined A.21 Windhover are promoted together in this advertisement dating from 1931. Offered for civil and military purposes, a total of twelve, twenty-one and two of each type respectively were built. All featured hulls constructed from ribbed 'Alclad'. (Aviation Ancestry)

Flying Boats
in production at the new works of
SAUNDERS-ROE LTD.
EAST COWES
ISLE OF WIGHT
Telephone : Cowes 393
Telegrams : Consuta, Cowes

By Appointment

After S.E. Saunders was taken over in 1929 by a consortium led by Alliott Verdon-Roe and the company was renamed Saunders-Roe, the development of an improved aircraft was begun, drawing on experience with the Severn. This was the A.27 London, a twin-engined military biplane. This advert from 10 December 1936 shows London aircraft under construction in the new, giant Columbine hangar opened in 1935. The view is through the large front doors (these days over-painted with the Union Jack) out into Cowes harbour. (Aviation Ancestry)

A successful design, orders for thirty-one Londons were secured from the Royal Air Force. Between April and June 1938, five of these aircraft from 204 Squadron completed an epic 30,000-mile flight from Plymouth to Australia and back. (Author's collection)

Another Saunders-Roe experimental aircraft was the solitary A.37 Shrimp completed as a scaled-down test bed for the planned Shorts' Shetland. This 1940 advert describes it as a 'Flying Miniature of a projected Ocean Air Liner', suggesting early on that the Shetland was intended for civil use as well as military duties. (Aviation Ancestry)

Above: The vast size of the Princess flying boat can be appreciated as the first craft, G-ALUN, is rolled out of the Columbine hangar on 30 October 1951. The fuselage length was 148ft with a wingspan of 219ft. Passenger capacity was for 100 persons on two decks. (Saunders-Roe)

Opposite top: The Shetland, one of the largest flying boats built up to that time, was contracted jointly with Saunders-Roe and Shorts at Rochester as a replacement for the famous Sunderland flying boat. The military version (S.35) was later complemented with a civil transport variant (S.40) – the aircraft G-AGVD shown here. The design was shared between the two companies while Saunders-Roe built the wings and the power plant assemblies. Only one of each type was built, the wartime acceleration of landplane development resulting in neither version of the Shetland being taken up for production orders. (Crown Copyright)

Opposite bottom: Despite the inability to secure orders for the Shetland, Saunders-Roe firmly believed there was still a future for large civilian flying boat transports. The outcome of its efforts, with BOAC long-distance services in mind, was the giant SR.45 Princess, the largest metal flying boat ever built. Here all three aircraft of the type are seen under construction in the Columbine hangar on 21 August 1950. Though it is a tight squeeze, it is also a good visual indicator of the great size of the building. (Saunders-Roe)

Only one Princess flew, the first aircraft to be completed. Here G-ALUN is seen lifting off the Solent after 'running high and proud' on its take-off run. (Saunders-Roe)

The magnificent Princess on flight trials over the Isle of Wight on 28 August 1952. Between August 1952 and June 1954 it accumulated 97 hours' flying time, including two appearances at the Farnborough Air Show. However, BOAC opted for landplanes for its long-haul routes and G-ALUN was decommissioned and cocooned at Cowes where she remained until July 1967. The two other Princess boats were cocooned at Calshot Spit in Southampton Water until they too were demolished in 1965. (Saunders-Roe)

3

STEAM POWER, ENGINES & PROPULSION

From 1830, when White's built its first steam-powered vessel, up to the late 1880s, the company had relied upon other British manufacturers to supply the engines to drive its ships, the principal source being the George Belliss firm in Birmingham. The shipping of engines to the Island for installation was, though, a far from ideal arrangement. Therefore, in the 1880s John Samuel White determined that it was time for his company to commence building engines itself along with all the associated steam machinery necessary to complete the entire power systems for the steamships of the time.

Consequently, the decision was made to expend with the remaining Medina dry dock located in West Cowes and utilise the reclaimed land for the erection of a large and comprehensive engineering works, commenced in 1889. When the works opened, among its senior staff from the outset was a talented design engineer, Andrew Forster, who, inspired by John Samuel White, set about the development of a superior ship's boiler.

Andrew Forster was born at Jarrow on Tyneside in 1867. He received his education at Rutherford College, Newcastle upon Tyne – which still specialises in Science, Technology, Engineering and Mathematics (STEM) subjects – following which he served his apprenticeship at Palmers Shipbuilding & Iron Company at Jarrow-on-Tyne. On its completion, in 1887, he was employed as a draughtsman by Robert Stephenson & Co., Newcastle upon Tyne, then Palmer & Co., followed by the Naval Construction & Armaments Company in Barrow-in-Furness, years in which he gained a depth of valuable design experience. In 1896 he was appointed chief draughtsman and manager's assistant at J. Samuel White, working in the Engineering Department. He was soon promoted to engineering manager. Besides the patented White-Forster water-tube boiler, designed at this time, he was also responsible for many other patented engineering designs including the White-Forster automatic feed water regulator, the 'dummyless' marine turbine, oil fuel sprayers and fuel injection devices and an improvement to calculating slide rules. From 1921 he engaged in a programme of improvements to semi-diesel engines, many of which were patented, and in 1922, with Canadian designer Arthur George Mellor, he invented a new type of internal combustion engine.

The White-Forster boiler, patented in September 1898, had been preceded by the invention of a novel circulating coil-type water-tube boiler suitable for installation in smaller craft. In the larger White-Forster three-drum boiler, Forster's design utilised curved water tubes to ensure secure jointing of the pipe ends and ease of maintenance. Rivalled only by the comparable Yarrow boiler, both of which could raise steam with far greater efficiency, temperature and pressure than the earlier Scotch boilers which were of the fire-tube configuration, the White-Forster boiler became a huge success. It was thenceforward installed in all the steamships built in the yard and subsequently adopted by other shipyards both at home and abroad, either supplied to them as complete installations or built by them under licence to White's.

Shortly after the opening of the Engineering Works, another influential person, Mr Edwin Charles Carnt, arrived on the scene to take over as engineering director and general manager. Edwin Carnt's background was in the Royal Navy. Born in Portsmouth in May 1858, he was educated at the Royal Naval College, Greenwich, and on his graduation he entered the Royal Navy as an assistant engineer serving aboard ships at sea. Subsequently promoted to the rank of engineer, he was later shore-based, transferred to Barrow-in-Furness as engineer overseer on behalf of the Admiralty. He was promoted to chief engineer in June 1893 and to staff engineer in June 1897. Less than a year later, in February 1898, he resigned his commission to take up his position at J. Samuel White.

During his seventeen years with the company Edwin Carnt instigated and supervised a major modernisation and expansion programme of both the shipbuilding and engineering infrastructure to equip the business for burgeoning naval orders as well as to meet the growing demand for White-Forster boilers and steam machinery. This included the refurbishment of the engine shops, the building of a new boiler shop in West Cowes, the extension of the foundry and the erection in 1912 of the 80-ton electric cantilever hammerhead crane. These developments required the reclamation of land on both sides of the river.

As a measure of the impact of this massive investment programme, employment in the Engineering Department grew from 350 in 1898 to 1,600 by 1915 and in the shipyard from 250 in 1898 to 1,500 by 1915. By the time of his death at Wootton, Isle of Wight, in August 1915, he had become chairman and managing director, the latter position held since 1911.

These advances also witnessed a broadening diversification in the output of engine products and, as a result, White's became known for its pioneering adoption of alternative forms of marine engine and propulsion systems. Despite its relatively small overall area, within which were eight main workshops plus foundries and tool stores, the Engineering Works rapidly moved on from compound and triple-expansion steam-reciprocating engines to steam turbines and, later, internal combustion engines. Not satisfied with simply working to the licensed designs developed elsewhere, the company's engineers embarked upon the enhancement of these systems wherever it saw room for improvement, building and testing full-scale experimental models which eventually saw service in vessels constructed in the shipyard. Besides prime moving machinery, White's collaborated with other concerns to produce diesel, kerosene and paraffin-fuelled generators of small and medium size for a range of applications.

Throughout the two world wars, the output of the Engineering Works at White's rivalled that of the main shipyard with countless boilers, diesel sets and turbine and reciprocating steam engines. Besides new machinery, a considerable amount of overhaul and repair work was also handled.

As it turned out, the Engineering Works out-survived the shipyard by some sixteen years after it had become a subsidiary of the American-owned Carrier Corporation, the last five under the name Elliott Turbo-Machinery. In fact, because of the departure of the last of the shipbuilding foremen and charge hands from the yard in April 1965, engineering staff were called upon to supervise the completion of the yard's last ship, the frigate HMS *Arethusa*.

Even in those threatening years of decline and closure, White's remained committed to its innovative traditions, launching a line of water-jet marine thrusters and pioneering a product type which, in recent years and in a slightly different form, has enjoyed massive sales: the water scooter.

As an indication of the growth of the J. Samuel White Engineering Department, its total annual output of 3,000shp in 1898 had increased to well over 100,000shp by 1915. In the First World War it grew even further, a trend that continued through the inter-war years and the Second World War.

Mention should also be made here of some of the other important and groundbreaking marine engine and engine control products which have emanated from local manufacturing concerns besides J. Samuel White. Set up in East Cowes in 1894 by the American businessman Henry Alonzo House with Robert Symon as his financial backer, the Liquid Fuel Engineering Company (LIFU) was among the first manufacturers to produce petrol-fuelled steam engines. Occupying the Columbine Yard premises, formerly run as a boatbuilding business by G. & A. Spencer, it was not long before LIFU had as many as 200 employees. Small by comparison to those produced across the River Medina at White's, LIFU's compound engines with their oil fuel-burning boilers were installed in a range of small marine craft and in a variety of road vehicles such as delivery vans and charabancs. Of interest, the Cheverton company in Newport made and bound the wheels for LIFU vehicles.

Unfortunately, the company was compelled to close its operation in East Cowes in 1900 partly because of difficulties in shipping its road vehicles across the Solent. The cost of cross-Solent transportation still remains a contentious issue today and, no doubt, it has had a bearing on the fortunes of other Island enterprises. It certainly influenced businesses like White's and, more recently, Vestas to provide their own means of conveyance to and from Southampton. As for LIFU, it relocated at Poole in Dorset and at Northam, Southampton, where it continued in production until around 1915 when the company folded completely.

In more recent times, Saunders-Roe developed the contactless marine torquemeter, a device using FM telemetry to measure and record the tendency to twist of a rotating object, such as a ship's propeller shaft or an engine transmission system, when a force is applied to it. The purpose is to optimise performance, so as to achieve greater efficiency while maintaining the required thrust, and reduce wear and tear of the system. One of the great benefits of the employment of these torquemeters is the savings in fuel consumption than can be achieved as a result. Utilising wire-strain gauges, also invented at Saunders-Roe, conventional torquemeters had originally been applied for other measurement purposes but the complicated power transmission arrangement of the engine pylons of the SR.N4 hovercraft presented a unique torque measurement difficulty requiring a contactless system. Conceived and developed by Eric Jolliffe at the Experimental & Electronics Laboratories (EEL), later renamed Test Facilities, the subsequent utilisation of this equipment aboard ships was an ideal solution for better ship-operating economics. Today the company's contactless torquemeters are installed on cruise ships and cargo vessels of all types, and also on naval submarines and aircraft carriers where it is now designated as the Standard Naval Torsionmeter. The torquemeter is an application of Hooke's Law, a principle of physics relating to the behaviour of elastic bodies when forces are applied to them. Robert Hooke, its originator, was also an inhabitant of the Isle of Wight.

As evidence of the innovative skills to be found on the Island, another recent example of a small but important engineering solution conceived to resolve an operational problem is a contactless pilot control, devised by two local electronics engineers from Carisbrooke Engineering, who prefer to remain incognito, for the operation of the swivelling puff ports of AP1-88 and British Hovercraft Technology (BHT) hovercraft. Mounted forward, puff ports are deployed to manage the lateral movement of these craft by rotating through 360 degrees, blowing air to provide directional control in addition to that available from the rudders within the main propulsion ducts. The need to engage these devices on a frequent basis to counter crosswinds has resulted in rapid wear to the existing contact-type pilot controls. The new, contactless controls, which utilise Reed switches activated by an applied magnetic field, have almost unlimited endurance. This has saved hovercraft operators endless costs of having to constantly fit replacement parts.

Above: The company sought constantly to make improvements to current engine types. In this case, an experimental reciprocating engine constructed to the designs of Henry James Heasman. (Isle of Wight Heritage Services)

Opposite: White's Engineering Works initially built compound (double-expansion) steam-reciprocating engines to Admiralty standard designs. Triple-expansion reciprocating engines then followed as the capabilities of the Engineering Department developed. Output was not only for the company's shipyard but also for builders elsewhere. This three-cylinder triple-expansion steam engine developing 1,400hp was destined for HMS *Cadmus*, a sloop built at the Sheerness Dockyard in 1903. (Debenham, courtesy of Isle of Wight Heritage Services)

Above: The White-Forster boiler was highly regarded for its superior performance and it was rapidly adopted for new steam vessels built at White's and made under licence by other shipyards. Its greatest tribute was its selection for the giant American aircraft carrier USS *Saratoga*, the world's largest warship between 1927 and 1940 at 49,552 tons and 909ft in length. She had sixteen White-Forster boilers supplying steam to twin turbo-electric engines. The complete power plant had an output of 180,000shp giving the carrier a top speed of 33 knots (38mph). (United States National Archives)

Opposite top: The Engineering Works greatest innovation was the patented White-Forster three-drum water-tube boiler designed by Andrew Forster. Similar to the Yarrow boiler, the all-riveted White-Forster boiler incorporated features for ease of maintenance. The water drums are at either side at the bottom were connected by the tubes to the steam drum at the top. The boiler shown here was for one of the Chilean Almirante Lynch-class destroyers. (Debenham, courtesy of Isle of Wight Heritage Services)

Opposite bottom: A completed White-Forster boiler enclosed in its outer casing. The doors to the furnaces can be seen centre bottom located between the two water drums. This boiler, like that in the previous photograph, was destined for an Almirante Lynch-class destroyer. (Isle of Wight Heritage Services)

White's continued to build White-Forster water-tube boilers for many years, over time encompassing many of Andrew Forster's patented improvements. Here, a White-Forster boiler is lowered by the hammerhead crane into the Royal Navy destroyer HMS *Havant* (ex-*Javary*) while fitting out in 1939. (White's archives, Isle of Wight Heritage Services)

Based in East Cowes, the Liquid Fuel Engineering Company (LIFU) was established in 1894 by American businessman Henry Alonzo House. The company specialised in small steam engines with boilers burning paraffin or petroleum rather than solid fuel at a time when most marine steam engines were still fuelled by coal. The photograph shows a typical small LIFU steam launch putting into West Cowes with the distinctive Coast Guard cottages in East Cowes in the background. (Kirk, courtesy of Isle of Wight Heritage Services)

LIFU's two-cylinder compound engines were not only fitted into marine craft but a range of road vehicles including delivery vans, as shown here, charabancs and motor cars. (Author's collection)

A somewhat larger LIFU-built vessel is this unidentified 75ft twin-screw steam launch that dates from around 1900. (Kirk, courtesy of Isle of Wight Heritage Services)

Demonstrating White's innovative spirit, before the First World War it had moved on to the production of internal combustion engines. Illustrated here is an improved, experimental, six-cylinder, two-stroke single-acting diesel engine, two of which were installed in the motor yacht *I Wonder*, built in 1912 for John Samuel White's eldest son, John Lee White. (Photo: Beken)

In conjunction with Crompton Parkinson, White's developed a 4kW diesel generator which employed a small, high-speed oil engine. The company also designed and produced a 'Super Semi-Diesel Engine'. (Isle of Wight Heritage Services)

As early as 1906 White's had begun constructing Parsons steam turbines under licence for the five Cricket-class coastal destroyers, later reclassified as torpedo boats. The work involved in manufacturing turbine blades, machining rotors and casting rotor casings, besides blading the rotors, all revealing something of the capabilities of the Engineering Works' fitters in working conditions that were not exactly conducive to such precision engineering. The photograph shows an early turbine under construction for the Tribal-class destroyer HMS *Mohawk*, completed in 1908. (J. Samuel White)

In this view, blading is in progress on the port turbine rotor for the *Caesarea*, one of two cross-Channel steamers built in 1960–61. Each of these ferries had twin 4,500shp steam turbines. During his time as Engineering Department manager, Andrew Forster had also introduced a series of patented improvements to steam turbines. (Ron Trowell)

This White-Brons paraffin-fuelled auxiliary services generator was manufactured for installation aboard the flotilla leader HMS *Lightfoot*, completed by White's in 1915. Small engine units of this type were fabricated in quantity for all the White's-built Royal Navy destroyers of the First World War period. (Photo: Beken)

Opposite top: The construction of gas turbines, gas expanders and turbo-compressors was a business that was to prove something of a lifeline to J. Samuel White after the shipyard closed down. This activity commenced with the manufacture of the gas turbine engine for the mixed-power Tribal-class frigate HMS *Eskimo*. The photograph shows the engine's half compressor cylinder. (Ron Trowell)

Opposite bottom: This view shows the bladed turbine rotor prior to enclosure within the complete compressor assembly. Delivered in 1963, HMS *Eskimo* had a Combined Steam and Gas (COSAG) power plant driving a single screw. (Ron Trowell)

Right: Like other local companies, White's also produced a wide variety of peripheral engine products and components for a diversity of purposes. One of these was its range of Clinsoll Strainers for engine oil filtration as shown in this advert which appeared in *Engineering* magazine in September 1966. (Isle of Wight Heritage Services)

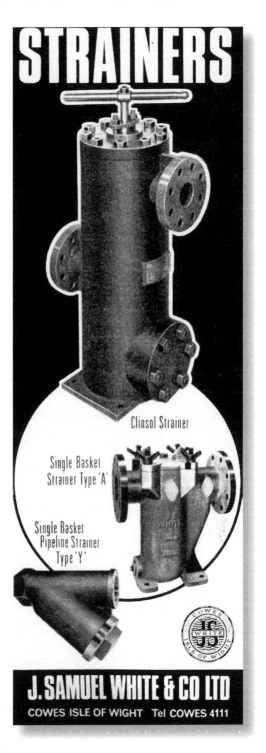

STRAINERS

Clinsol Strainer

Single Basket Strainer Type 'A'

Single Basket Pipeline Strainer Type 'Y'

J. SAMUEL WHITE & CO LTD

COWES ISLE OF WIGHT   Tel COWES 4111

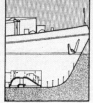

Opposite top: Two of White's female employees are seen here assembling Clinsoll Strainers in the early 1960s. (Brian Greening collection)

Opposite bottom: Another important marine engineering innovation from J. Samuel White was the propellerless marine jet thruster which had been under development since 1922 for inland waterway use. Its full potential became apparent following the installation of the first deep-sea thruster unit in 1966. Several versions of marketable thrusters were produced either as the stern-mounted main propulsion system on such craft as fast ferries, obviating the need for a rudder, or fitted athwartships for manoeuvring or station keeping. The advert, printed in *Ship & Boat Builder International*, dates from July 1967. (Isle of Wight Heritage Services)

Below: The White-Gill thruster shown here during manufacture at West Cowes is of the hull-mounted. Production of marine thrusters continued after White's was taken over by Elliott's (Carrier Corporation) in 1971 but, sadly, despite lucrative orders and the need to retain local employment, production was subcontracted to Tees Components located at Saltburn. Subsequently, the entire business, design specifications and customer accounts were sold to the Teeside company. Production still continues there under the same product name and a unit of the type illustrated was fitted to the research ship RRS *Discovery* as recently as 2013. (Ron Trowell)

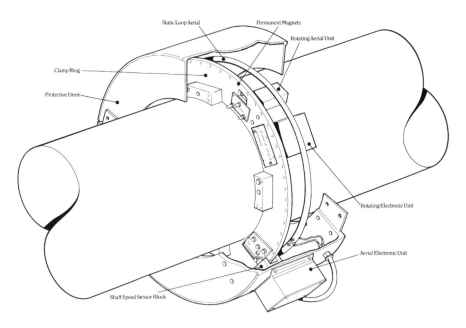

Saunders-Roe's Experimental & Electronics Laboratories (EEL) subsidiary pioneered strain gauge, load cell and torque measurement technology. Having developed a range of torque measuring devices, EEL next went on to produce the world's first contactless torquemeter or torsionmeter, because instrumentation attached to a shaft could induce vibration, affecting signal quality. As shown in this diagram, the contactless torquemeter utilises FM telemetry and a loop aerial to supply power to the installation and to transmit data back to the reading device. The concept, the brainchild of designer Eric Jolliffe, can support multiple applications. (EEL Limited)

By determining the horsepower requirements to satisfy a particular work function, torquemeters allow operators to optimise control of the power output onto a shaft. One of the main applications of the contactless torquemeter was for ships' propeller shafts as shown here of this installation aboard the aircraft carrier HMS *Illustrious*. The circular drum structure surrounding the shaft contains the static loop aerial. Also within it are the rotating aerials attached to the surface of the shaft. (John Smythers)

4

THE QUEST FOR SPEED
OVER THE WATER

Boatbuilding on the Isle of Wight has been concentrated in two main areas: along the east and west banks of the River Medina and around Bembridge Harbour at St Helens and Bembridge. Smaller pockets of activity are also located at Yarmouth and Wootton. Particularly numerous were the boatyards that were once to be found along either shore of the Medina in West and East Cowes. Mention has already been made of Saunders' Cornubia Yard in East Cowes. Besides that, there were the Sunbeam, Goshawk, Sylvia, Minerva and Marvins boatyards while on the west bank there were, among others, the Vectis Works, Lallows, Victory Yard, Whitegates Yard and a second yard owned by George Marvin. Collectively, over the years these many boatbuilders have accounted for a major part of the Island's industrial output, producing everything from yachts, motorboats and naval pinnaces to rescue launches, lifeboats and workboats. They have also produced some of the most exciting and successful high-performance racing craft.

As mentioned in the preamble to Chapter Two, Samuel Saunders boatbuilding concern had soon acquired a reputation for the construction of race-winning high-speed marine craft, establishing a pedigree with boats like the *Yarrow Napier*, winner of the Harmsworth Trophy in 1906, and *Ursula*, which took the world's fastest vessel title in 1910 at 37.9 knots (43.6mph). This culminated in the *Maple Leaf IV*, designed by Joe Porter, a five-stepped hydroplane which claimed the world water speed record on 11 September 1913 at 50 knots (57.5mph). The *Maple Leaf V*, *VI* and *VII* which followed did not quite gain the acclaim of their famous antecedent, although *Maple Leaf VII* achieved a maximum trials speed of 69.5 knots (80.0mph).

In the 1920s, after employing other designers following the departure of Joe Porter to Groves and Guttridge, Saunders continued with yet more racing hydroplanes and hard-chine boats, including a Puma series inspired by Hubert Scott-Paine and the *Estelle I* and *Estelle II* designed by Percy Hyde Beadle and constructed for wealthy pioneering lady racer Marion 'Betty' Carstairs. When Miss Carstairs most impressively determined to embark upon the unprecedented challenge of crossing the Atlantic at record speed, Hyde-Beadle also designed for her a very large 78ft multi-stepped hydroplane. Despite its size and unique hull form as well as its ability to reach speeds of more than 43.5 knots (50.0mph), this amazing craft, envisaged well before its time, did not cope well with the heavy swells of the unsheltered ocean and, unfortunately, the attempt had to be abandoned. As a matter of interest, Sir Richard Branson finally accomplished this feat over half a century later when he took the non-commercial Atlantic speed record on 29 June 1986, crossing in three days and 8½ hours at an average speed of 36.7 knots (42.1mph), an achievement which also had a connection with the Isle of Wight. Although his boat, the 72ft delta-hulled *Virgin Atlantic Challenger II*, was not constructed on the Island, it did come from the drawing board of Island-based designer Renato 'Sonny' Levi.

While the large transatlantic craft built by Saunders for Betty Carstairs did not perform as well as had been expected it did not deter Saunders from continued efforts with high-speed racing craft. However, Betty Carstairs went on to acquire the vacant Sylvia Yard where she had the next two *Estelle* craft built by her own team. Sadly, the British International Harmsworth Trophy, a prize she had committed herself to winning, constantly eluded her despite the fortune she had spent on its pursuit. Thereafter she lent her support to the record-breaking attempts of other drivers, notably Sir Malcolm Campbell.

Besides Percy Hyde Beadle, who was dismissed following the failure of his Atlantic boat, the Saunders' design team had been complemented in the early 1920s by the services of a truly notable boat designer, Fred Cooper, who went on to design two superior world speed record-breakers, both built at East Cowes.

Fred Cooper was born in November 1898, the youngest son of Frank Cooper, a monumental mason in Newport, Isle of Wight. He was educated locally but whereas his two older brothers entered his father's business he aspired instead to become a naval architect, a goal he achieved as a hull designer of exceptional capability. He worked for Hubert Scott-Paine, founder of both the British Powerboat Company at Hythe and the Supermarine

Aviation Company which produced the Schneider Trophy-winning aircraft as well as, most significantly, the Spitfire fighter plane. Around 1924, Fred Cooper joined S.E. Saunders as chief draughtsman at the Cornubia boatyard, among his creations for the company being the racing powerboat *Newg* for Marion Carstairs and the Puma class of four racing cruisers capable of 41 knots (47.0mph).

He did not, however, remain employed by the company for long. Preferring the independence to pursue his own commissions, within five years he had left to become a self-employed naval architect based at Hythe gaining for himself an enviable reputation. But when he was contracted to design what became his two greatest powerboat achievements, his old firm was also commissioned to undertake their construction. Having already designed *Miss England* for Sir Henry Segrave, built by the British Powerboat Company, he went on to design *Miss England II*, also for Segrave, and *Bluebird* K3, for Sir Malcolm Campbell, both of which gained the world water speed record. It was said of the latter boat that 'In its day, it was the most advanced craft of its type in the world'. Fred Cooper also carried out the preliminary design work for the three-point hydroplane *Bluebird* K4, another world water speed record holder. Described as talented and prolific, coming from a family of skilled men, it is considered nonetheless that he never fully received the recognition he deserved in the way that his modern-day equivalent would have.

The craze for motorboat racing and breaking the world water speed record led on to the pursuit of high-speed craft for other, more practical, applications, a matter of increasing importance as another world war loomed. Many years later, the focus for this type of craft would be on fast, short-route ferries but in the late 1930s the concern was to evolve lightweight, high-speed naval craft for coastal operations. The experience gained with racing boats and record-breaking hydroplanes was invaluable for this effort. One project of particular interest involved the shipbuilders J. Samuel White which hitherto had not engaged in the development of such craft. This was all to change when a concept for a hydrofoil ship was patented in June 1935 (patent No. 458,111), subsequently fashioned into a scaled-down trials craft built in 1937 under the appropriately secretive project name 'Hush-Hush'. Having gained the attention of the Admiralty, this in turn resulted in a private venture full-size aluminium-skinned vessel, yard number 1808 subsequently designated MTB 101. This vessel, capable of a speed of 42 knots (48.3mph), was almost certainly one of the first full-scale craft of this type to be built in the United Kingdom.

During the ensuing war, most of the local boatbuilders produced fast craft for defence purposes. White's built some forty motor torpedo boats (MTB) and motor gun boats (MGB) to its own designs for the Royal Navy and Polish Navy. White's also produced Air Sea Rescue launches as did other boatyards like Saunders-Roe and Groves & Guttridge, while Woodnutts at Bembridge turned out twenty-four Fairmile fast motor launches (ML) for the war effort.

At the war's end, the initiative with high-speed marine propulsion switched to Saunders-Roe which produced the first aluminium-hulled displacement MTB for the Royal Navy, reviving the basic design for such a proposed vessel which had been first conceived by the company back in 1932. This very successful craft, which had a maximum speed of 42 knots (48.3mph), became the prototype for the subsequent Dark class of fast patrol boats, of which six boats, like the prototype MTB 1602, were built by Saunders-Roe at its Beaumaris premises in Anglesey, North Wales, a dispersal plant that had been opened there during the war.

A later project, designed at East Cowes and built at Beaumaris, was the 17-ton hydrofoil *Bras D'Or*, completed for the Royal Canadian Navy. Around the same time, the inventor and designer Christopher Hook was working on his experimental hydrofin craft *Icarus*, distinctive for having mechanically connected projecting 'feeler' or 'jockey' floats which sensed the approaching wave height to adjust the inclination of the lifting foils. It could be seen throughout 1952 and 1953 running trials in Cowes harbour (see British Pathé film, reference 1283.01).

Competitive motorboat racing was revived in the early 1960s when Sir Max Aitken was instrumental in organising the annual Cowes to Torquay offshore powerboat race, later known as the Cowes Classic. Within just a few years this proved to be the stage upon which two distinguished designers and two notable local boatbuilders, Wilf Souter's yard at West Cowes and Enfield Marine at Fishbourne, were to make their names by introducing advanced racing craft that set the trend for years that followed. Of the two designers, one was Renato 'Sonny' Levi who designed the revolutionary and legendary *Surfury* as well as *Merry-Go-Round*, the winner of the world diesel-powered water speed record, and *Avenger Too*, the winner of the first Round Britain Race. The other was American designer Don Shead who produced a sequence of Cowes Classic winners such as *Telstar, Miss Enfield II* and *Enfield Avenger*. The latter, under the new name *Unowot*, had raised the average speed for the 200-mile course by 1975 to the then unheard of 63.4 knots (72.9mph).

The extraordinarily graceful *Surfury*, probably the most admired of classic racing powerboats, even fifty years after she was completed, has been preserved, donated to the nation and is now under the custodianship of the National Maritime Museum. Her designer, Renato 'Sonny' Levi, an extraordinary design engineer, now resides on the Isle of Wight where he based his business.

Born in Karachi in 1926 to Italian parents, Renato Levi was educated in India, France and England before joining the RAF during the Second World War reaching the rank of flying officer. On demobilisation in 1948 he studied aeronautics and aircraft design in England before returning to Bombay to work in the shipyard his father had set up building boats under contract to the government. By 1950 Renato held the position of chief designer in his father's drawing office. Ten years later he founded the Canav Navaltecnica company in Anzio, designing and building high-speed pleasure boats, the precursors to the modern planing boats incorporating the deep-V hull form which became Renato Levi's hallmark.

He established his name with striking racing boats beginning in 1961 with *A Speranziella*, his first complete design for a fast cruiser, built from a combination of glass reinforced plastic (GRP) and exotic woods. Renato Levi's famous Delta line of needle-nosed deep-V planing boats was established in January 1965 with the first of the breed *Surfury*, soon followed by *Merry-Go-Round*, *Delta Synthesis*, *Delta Blu* and other extraordinary streamlined racing boats. A derivative of this hull form was employed for Sir Richard Branson's Atlantic record-breaker *Virgin Atlantic Challenger II*. Apart from offshore racers, Renato Levi designed luxury craft, fast commuter boats, workboats and naval patrol vessels. Not only responsible for the deep-V delta hull exhibited in the *Surfury* and other contemporary craft he also created his patented 'power trim' surface-drive propulsion system.

In 1987, Renato Levi was made Royal Designer for Industry, an award to designers who have achieved 'sustained design excellence, work of aesthetic value and significant benefit to society'. Mike James of *Classic Offshore* said of Renato Levi, 'When the talk is of modern day powerboat racing, there are two names which will be forever linked to its history and spoken of in god-like esteem: Levi and *Surfury*.'

Clearly, the spirit of innovation and competitive inspiration is far from dead on the Isle of Wight as another recent and quite incredible, and totally successful, marine speed achievement demonstrates. Under designer and project leader Malcolm Barnsley, the *Vestas Sailrocket* was conceived in a bid to take the world sailing speed. When the first *Vestas Sailrocket* craft failed in its challenge, the team set about developing an improved version drawing on the experience that had been gained with the first craft. Constructed in the Medina Shop alongside the famous Columbine hangar from which so many other revolutionary and extraordinary craft have emerged, the *Vestas Sailrocket 2* was completed in March 2011. Taken to Walvis Bay, Namibia, it achieved the phenomenal speed of 65.45 knots (75.3mph) on 24 November 2012, piloted by Paul Larsen, taking the world 500m sailing speed record. Its peak speed of 68.33 knots (78.5mph) is only acknowledged unofficially but it also smashed the nautical mile world record with a speed of 55.32 knots (63.6mph).

The spirit of the quest for speed over the water by small teams and local businesses was most definitely not exhausted. It moved next from hydroplanes and powerboats, which exploit developed forms of hull geometry to raise the craft up onto the water surface to minimise drag, to another more radical approach to improved hydrodynamics: the hovercraft, in which a layer or cushion of low pressure air lifts the craft above the sea, virtually eliminating water resistance. These craft are dealt with in a later chapter but suffice to say here that the Isle of Wight became the birthplace of the hovercraft and within a period of just eight years had not only launched a production line of commercial craft but had produced the largest hovercraft ever built.

As a footnote to this preamble, as recently as July 2015, as a further endorsement of the Island's tradition in the construction of high-speed marine craft, an agreement was reached between South Boats of Cowes and Tampa Yacht Manufacturing of Florida, USA, whereby the Cowes yard will commence building fast patrol boats to American design in both aluminium and GRP for anti-terrorist and offshore exploration roles.

Although it is primarily promoting the use of patented Consuta sewn plywood for use on marine aircraft, this S.E. Saunders advert from 1924 stresses that it is the 'lightest and strongest material yet evolved for Marine and Aircraft Construction'. (Aviation Ancestry)

One of several record-breaking boats built by Saunders and Saunders-Roe fabricated from Consuta, the hydroplane *Maple Leaf IV* was adapted by Sidney 'Joe' Porter from a Henri Fabre design. *Maple Leaf IV* secured the world water speed record in July 1912 with Tommy Sopwith at the helm achieving 40.5 knots (46.5mph) over the measured mile and winning the British International (Harmsworth) Trophy. Her best speed was over 50 knots (57.5mph). Between 1920 and 1921, Saunders completed the successive craft *Maple Leaf V, VI* and *VII*. Though these racing craft were not as successful in competitions, *Maple Leaf V* broke the European water speed record at 57.6 knots (66.2mph) and *Maple Leaf VII* achieved a top speed in excess of 80 knots (92mph). (Kirk, courtesy of Isle of Wight Heritage Services)

Above: Having failed in her bid to win the British International Trophy, Betty Carstairs set her sights on a much bolder challenge: to break the transatlantic speed record then held by the ocean liner *Mauretania* at 26.2 knots (30.1mph). For the record attempt, Saunders built the impressive 78ft four-step hydroplane *Jack Stripes*, its name derived from the Union Jack and Stars and Stripes flags. The boat had a distinctive turtle back foredeck for water dispersal but, despite this and other innovations, the multi-step configuration did not cope well with the larger waves of the open ocean. The photograph shows *Jack Stripes* as converted to the cruiser *Voodoo* following the abandonment of the Atlantic challenge. (Photo: Beken)

Opposite top: The extraordinary craft shown here at the Folly Works in 1911, described as an aero-hydroplane, was designed by Roger Ravaud and built of Consuta by Saunders. Anticipated to be capable of high speeds, it was in effect an early attempt at a hydrofoil powered by an airscrew. Although the Ravaud aero-hydroplane did not apparently perform as calculated, it should not be dismissed as an impractical concoction – just nine years later Alexander Graham Bell constructed a similar craft, the *Hydrodome* HD-4, which broke the world water speed record at 61.7 knots (70.9mph). (Saunders-Roe)

Opposite bottom: Saunders had been commissioned to build three racing boats for the famed lady racer Miss Marion 'Betty' Carstairs who was determined to win the British International Trophy. All were designed by Francis Percy Hyde Beadle but the first two, the *Estelle I* and *II*, which were essentially trials craft for the design of the third, proved to be a disappointment. Consequently, Miss Carstairs acquired the Sylvia Yard in East Cowes, engaged designer Bert Hawker, and with her own team of boatbuilders constructed the *Estelle III* and *IV* herself, the latter shown here. Although *Estelle IV* achieved a top speed of 69.6 knots (80mph) neither boat met expectations and she failed to win the coveted Harmsworth Trophy. (Isle of Wight Heritage Services)

A more successful record-breaking powerboat completed by S.E. Saunders for Sir Henry Segrave was the 36ft *Miss England II* designed by Fred Cooper. With Sir Henry Segrave at the controls, she raised the world water speed record to 85.9 knots (98.8mph) at Lake Windermere on 13 June 1930, a feat which cost Segrave his life. The photograph shows *Miss England II* flat out at high speed. (Author's collection)

Sir Henry Segrave stands on the foredeck of *Miss England II* after his first attempt at the speed record on 5 June 1930. Later, after Sir Henry Segrave's death, *Miss England II* raised the world water speed record on three more occasions driven by Kaye Don, ultimately taking it to 95.9 knots (110.2mph) on Lake Garda on 9 July 1931. The only physical reminder of this acclaimed record-breaking craft is a model of the *Miss England II* in the Science Museum, London. (Brian Greening collection)

The 1930s was a golden era of racing and record breaking on land, in the air and on the water. One of Britain's best-known speed challengers, Sir Malcolm Campbell, commissioned Saunders-Roe to build his first *Bluebird* racing hydroplane, the K3, again designed by Fred Cooper. *Bluebird* K3 raised the water speed record to 109.9 knots (126.3mph) on 1 September 1937 on Lake Maggiore, Italy, increasing it slightly the following day. A year later, on 17 August 1938, Campbell raised his record speed to 113.9 knots (130.9mph) on the Hallwilersee, Switzerland. (Saunders-Roe)

Having reached the limits of this craft, Sir Malcolm Campbell turned next to Vospers for a three-point hydroplane designated K4 which used the same engine. Again Fred Cooper was involved in its design. *Bluebird* K3 has been fully restored and is once more in operating condition. She is seen here in October 2012 making the second of her post-restoration runs on Bewl Water, on the boundary between East Sussex and Kent. (Lisa Beaney Photography)

In 1932, Saunders-Roe designed an aluminium-hulled motor torpedo boat (MTB) for operation by the Royal Navy. With a length of 75ft and powered by three 600hp petrol engines for a projected speed of 30 knots (34.5mph), it was a portent of the future. Radical for its time, nothing more became of it. (Saunders-Roe, courtesy of Ray Wheeler)

Sixteen years later, the concept was revived and evolved to become the first all-aluminium displacement MTB for the Royal Navy, MTB 1602. Designed at Cowes and built by the Saro wartime dispersal plant established at Fryars Bay, Beaumaris, it became the prototype of the subsequent Dark class of fast patrol boats (FPB), six of which were also built at Beaumaris. Commissioned in 1949, MTB 1602 had the same length as the 1932 project but three 1,670hp Packard engines gave it a top speed of 41.8 knots (48mph). (Author's collection)

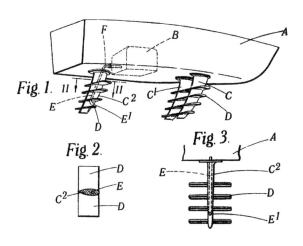

In the mid-1930s, J. Samuel White also began the development of a comparable project for an all-aluminium hydrofoil MTB for service use. On 13 June 1935, the company registered patent No. 458,111 for a hydrofoil craft with three foil struts, two forward and one aft. Shortly afterwards, an 18ft hydrofoil test craft was built under the project name 'Hush-Hush'. During trials the experimental craft achieved a maximum speed of 33 knots (37.9mph) driven by a 130hp engine. The drawings show the craft's general outlines. (Espacenet Patent Office)

The next stage of the development was the construction of a full-scale craft. Fabricated from 'Birmabright', the pioneering hydrofoil MTB 101, completed in 1939, was among the outstanding boats built by White's in the twentieth century. With ladder-type foils and powered by three Isotta Fraschini petrol engines developing 3,450bhp it was capable of rapid acceleration and a maximum speed of 42 knots (48.3mph). There is no doubt that the research programme was carried out in conjunction with the Admiralty which eventually acquired the craft for £51,000 but the records suggest that White's built the craft as a private venture. Despite its potential, it was vulnerable to foil damage under service operating conditions and it was lost altogether when one of the foils sheered off. The photograph shows MTB 101 during sea trials in the Solent. (Photo: Beken)

Designed as a mini-destroyer, the steam gunboat *Grey Goose* was commanded in the Second World War by Sir Peter Scott, participating notably in the Dieppe Raid in August 1942. The *Grey Goose* displaced 165 tons, had an overall length of 145ft, a beam of 20ft and a speed of 35 knots (40.2mph). After the war it was adapted as a test bed for gas turbine propulsion, as shown here. Still in existence, *Grey Goose* survives as the converted houseboat *Anserava*, moored on the River Medway at Hoo, near Chatham. (Author's collection)

The J. Samuel White-built MTB illustrated here powering along at a top speed of 40 knots (46mph) is the S7, one of six 73ft 47-ton boats completed in 1944 and transferred to the Free Polish Forces for operation as part of the Allied Coastal Forces Command. These craft had three Sterling petrol engines rated at 3,360shp. MTB S7 survives as the houseboat *Thanet*. (Maciej Bochenski)

While White's was constructing MTBs and MGBs, Groves & Guttridge were active building fast Air Sea Rescue (ASR) pinnaces and other types of fast craft at the Clarence Shipyard. This work continued well into the 1950s with new classes of boat of which the Vosper-designed 68ft ASR 2767, shown here, was completed in 1957. With twin Rolls-Royce Sea Griffon engines, her top speed was 39 knots (44.8mph). In total Groves & Guttridge built forty-nine RAF pinnaces of different classes. (Joe Thomas, Air Sea Rescue Marine Section Club)

Experimentation with gas turbine power plants continued into the 1950s, in this case in parallel with trials of different hull configurations. White's was commissioned along with Vospers to produce two almost identical high-speed mixed-power fast patrol boats, one each, fitted with twin Metro-Vick G2 turbines supplemented by twin Napier Deltic diesels for cruising. *Bold Pioneer*, the White's-built craft, proved the value of the hard-chine hull design over the rounded bilge configuration of the *Bold Pathfinder*, although their performance was comparable. Both vessels may be regarded as full-scale prototypes for subsequent vessels of this type. (Keith MacDonald)

In September 1953, Saunders-Roe was contracted by the Canadian Defence Department to produce the hydrofoil *Bras d'Or*, launched in May 1957. Designed and partly trialled in the Solent, the hull was constructed at Beaumaris. The foils and ancillary equipment were manufactured at East Cowes and transported to North Wales for assembly. The *Bras d'Or* achieved a top speed of 50 knots (57.5mph) during trials. Renamed *Baddeck* in 1962, she was decommissioned in 1973 and remains in storage at the Canada Science & Technology Museum in Ottawa. (Saunders-Roe)

The resumption of offshore powerboat racing in the early 1960s became the catalyst for some spectacular craft as designers strove to produce ever-faster boats. Probably the most notable of these and certainly the best remembered was the sleek, deep-V delta planing boat *Surfury* completed in 1965, designed by Renato 'Sonny' Levi and built by Wilf Souter at his boatyard in Cowes. The elegant lines of *Surfury*'s 40ft hull can be fully appreciated in this view of her on the launch cradle at Souters when newly completed. (Graham Stevens)

Underway at speed, *Surfury* made an impressive sight. Driven by the Gardner brothers, she won the Classic Cowes–Torquay Race in 1967 at an average speed of 46 knots (53mph) over the 198-mile course. Powered by twin Daytona 1,050hp engines driving a single propeller shaft through a Levi patented stern drive, she was capable of a top speed well in excess of 65 knots (75mph). Donated to the nation at the end of her racing career, *Surfury* is currently in the protective custody of the National Maritime Museum at Wroughton, Wiltshire, but it is possible that, at some future date, she may be returned to the Isle of Wight from where she originated. (Graham Stevens)

The diesel-powered *Merry Go Round* built by Wilf Souter in 1966 for Sir Max Aitken was a near sister vessel of the *Surfury*. She was cold-moulded in the hull plug that had been used for *Surfury* and thus shared the same dimensions and lines, although, unlike *Surfury*, she never had her covered cockpit fitted. Her engines were twin turbocharged VT8 Cummins diesels producing 1,100bhp. In March 1966, *Merry Go Round* broke the world diesel-powered water speed record at a speed of 52.4 knots (60.21mph). Renamed *Thunderfish III* a year later, as shown here, with 1,200hp Daytona petrol engines in place of the diesels, she blew up and caught fire during the Cowes–Torquay event of that year, sinking in Sandown Bay. (Graham Stevens)

With the break-up of the old established Royal Enfield Company, two of the constituent parts of the business were acquired and relocated to the Isle of Wight: Enfield Automotive, of which more later, and Enfield Marine, located at Fishbourne. The latter company, working with the American designer Don Shead, produced another series of exceptional race-winning powerboats along with others built at Souters. The *Miss Enfield II*, shown here, took the Cowes–Torquay Classic race in 1970 at an average speed of 50.9 knots (58.5mph) driven by Tom Sopwith, Charles de Selincourt and Don Shead. (Graham Stevens)

Miss Enfield II was followed by the 37ft *Enfield Avenger*, later renamed *Unowot*, as shown here. She too won the Cowes–Torquay race in 1973 at an average speed of 54.1 knots (62.2mph) over 255 miles. She was powered by twin Mercruisers rated at 1,000hp. Two years later, under the name *Unowot* she took the Cowes Classic honours for the second time. Re-engined with two Kiekhafer engines totalling 1,200hp output, she raised her winning speed to 63.4 knots (72.9mph) over 230 miles. The driver on that occasion was her designer, Don Shead, supported by Harry Hyams. (Graham Stevens)

In the shape of a quite different form of speed on the water, J. Samuel White pre-empted the later appeal of the personal thrills and spills of jet skis with its GRP 'Ski-ter' water scooters, launched in the 1960s. Powered by British Anzani Uniwin outboard motors, their exposed propellers perhaps presented a danger to other water users in the crowded beaches of resorts. Here one is being put through its paces in the River Medina, the rider requiring warm, protective clothing for the decidedly cooler climate of the UK. (Ron Trowell)

Another local company, Wight Plastics, later launched its improved version of the water scooter, the *Dolphin*, here on display with some of the design and production team. As far is known neither of these water scooters achieved volume sales. (Isle of Wight Heritage Services)

The *Vestas Sailrocket* 2 photographed while breaking the world sailing speed record at Walvis Bay, Namibia, on 24 November 2012. Exploiting the basic principles of aerodynamics and hydrodynamics, its design features a 'super-cavitating' carbon-fibre foil built by Independent Composites using an innovative heated vacuum-infusion method. The hull is formed of Gurit pre-impregnated carbon fibre with a Nomex honeycomb core and the wing skins are made of polyester heat-shrink film attached to a composite carbon spar. The craft's dimensions are 40ft length and 40ft overall beam with a total wing sail area of 22 square metres. (Helena Darvelid, Vestas SailRocket)

5

LIFE-SAVING, SPECIAL OPERATIONS & OTHER MARINE CRAFT

Besides producing craft for racing, record-breaking, leisure and competitive sailing, Island boatyards have also built many other boats for diverse practical and special purposes, once more demonstrating that necessity is the mother of invention.

As far back as the 1840s, prompted by a request from the P&O Steam Navigation Company (the forebears of today's P&O Cruises), John White and Andrew Lamb developed their patented lifeboat of which, consequentially, in excess of 800 were built and possibly as many as 1,000. Conceived originally as a shipboard lifeboat, it was not a self-righting boat, but because of the extraordinary buoyancy built into the hulls it was virtually impossible to capsize them. Many shipping lines purchased these lifeboats as word of their qualities became widespread, and hundreds were supplied to the Admiralty. Considered to be equally ideal for shore rescue purposes, a great many were acquired by independent lifeboat stations, the first delivered to the St Dogmeals station, Cardiganshire in 1849. They were not, however, adopted by the National Institution for the Preservation of Life from Shipwreck (later the RNLI), which had been founded in 1824, principally because it was not a self-righter, yet many stations subsequently taken over by the institution insisted on keeping their Lamb & White boats.

Many years later, the Isle of Wight firm of British Hovercraft Corporation (BHC), formerly Saunders-Roe, came to the aid of the RNLI by conceiving a self-righting solution to enhance certain of the non-self-righting lifeboats of the Watson and Barnett classes. Following the tragic disaster to the crew of the J. Samuel White-built Watson-class boat *TGB* from Longhope station on 17 March 1969, a group of designers led by Stuart Welford developed a modification kit at BHC's EEL division which, on a one-off basis, would right an overturned boat which otherwise had no means of righting itself.

Following the Lamb & White patented lifeboats, local boatbuilders went on to construct a total of 419 lifeboats of RNLI design, a record for commercial firms collectively in a single area of the country. It is quite possible too that J. Samuel White's holds the record for producing the most lifeboats for the RNLI by a single private company. The following list provides a breakdown of the output of Island boatyards, all based at West and East Cowes, in respect of RNLI lifeboat construction:

Builder	Years Active	Number
Hansen	1887–96	17
J. Samuel White	1898–1964	134
Saunders/Saunders-Roe	1915–37	61
Groves & Guttridge	1931–75	119
Fairey Marine	1980–88	27
FBM	1989–2000	32
Souters	1994–2002	29

Another vitally important life-saving craft which originated on the Island should also be acknowledged here. This is the airborne lifeboat designed by famed Isle of Wight yacht and boat designer Uffa Fox. Constructed at Saunders-Roe which received many orders for these craft, it was deployed in quantities and in different marks on Avro Lancaster, Vickers Warwick and Lockheed Hudson bombers as well as other planes during the Second World War. Post-war it continued in use on the Avro Shackleton oceanic patrol aircraft.

Another local man, Uffa Fox was born on the Island in January 1898, raised in East Cowes and attended school at Whippingham. He served an apprenticeship with S. Saunders & Company and, during the First World War, joined the Royal Naval Air Service, stationed at Felixstowe. Over the ensuing years, he pursued a career in sailing, designing and building yachts and boats. He was responsible for numerous developments in boat design, particularly of smaller sailing craft, which contributed to the growth in popularity of sailing as a sport and leisure activity. Running his own business initially from a former floating bridge, adapted as his design office and workshop and complete with living quarters, many classes of sailing boat were designed by Uffa Fox. Among them were the International 14, Foxcub, Super Foxcub, Flying Ten, National 12, National 18, Albacore, Firefly and, of course, the Flying Fifteen of which class the *Coweslip* was donated as a wedding present from the people of Cowes to Princess Elizabeth and Prince Philip on the occasion of their wedding on 20 November 1947.

Significant to this volume, besides his design for the airborne lifeboat that saved the lives of countless aircrews who were forced to ditch into the sea, two record-breaking endurance rowing boats also had their genesis on Uffa Fox's drawing board. They were the *Britannia* and *Britannia II*, which were built in West Cowes by Clare Lallow. In the first of these boats, John Fairfax established a solo Atlantic rowing record, a feat never before achieved, after setting out on his intrepid mission from the Canary Islands and making his landfall in Florida 180 days later, on 19 July 1969. Today, the 22ft-long, bright orange *Britannia* is an exhibit at the Classic Boat Museum, West Cowes. With companion Sylvia Cook, John Fairfax then set about completing the first rowed crossing of the Pacific Ocean in *Britannia II* between 26 April 1971 and 22 April 1972, just four days short of a year. Made from cold-moulded mahogany, these partially covered, self-righting and self-bailing craft were described as 'the Rolls-Royces of rowboats'.

Over and above his huge contributions to boat design, Uffa Fox was an eccentric, a wit and something of a philosopher. His larger-than-life, inspirational personality did much to draw world attention to the local yachting and boatbuilding scene at Cowes.

A quite different maritime life-saving appliance devised by White's in the 1800s for shipboard use was the patented Hire & White Lifebridge, a buoyant structure that was attached on runners above the bridges of steamers of the day, permitting it to be transferred to either side of the ship in an abandonment emergency. By so doing it ensured that a substantial part of a ship's life-saving apparatus was not lost because of an extreme list to one side. The Admiralty procured the lifebridge for installation on many of its transports of that period, including the HMT *Himalaya* and *Orontes*.

Before leaving the subject of lifeboats, mention should be made here of the continuing production of life-saving craft at the RNLI's own Inshore Lifeboat Centre in East Cowes. Rigid Inflatable Boats (RIBs) of the Atlantic class are produced there in quantity as well as maintained and overhauled.

Driven by twin 115hp four-stroke Yamaha outboards, the 27.8ft Atlantic 85 class have a top speed of 34.8 knots (40mph), while the smaller 24.3ft Atlantic 75 class they have replaced can reach 32.2 knots (37mph) driven by twin 75hp two-stroke Yamaha engines. Fifty of these boats and smaller inflatables are produced at East Cowes each year, constituting 60 per cent of the total requirements of the institution for this type of craft.

Many other novel craft for special purposes have been devised by Island designers, at least two of them the brainchild of the gifted boatbuilder Fred Goatley, employed at Saunders-Roe. Fred Goatley hailed from Binsey, Oxfordshire, where he was born in July 1878. A respected and skilled boatbuilder, his association with Sam Saunders began in the Saunders boatyard on the River Thames at Goring and Fred was one of a group of senior craftsmen who moved to the Isle of Wight in the early 1900s when the Saunders' Patent Launch Building Syndicate opened new premises in West Cowes. Fred Goatley was also an inspired and innovative designer who excelled with concepts for folding, sectional and collapsible boats to meet a variety of needs. By 1937, he was a manager at the Consuta stitched-ply production facility at Whippingham, subsequently named Saro Laminated Wood Products.

In the years preceding the Second World War he designed a standard canoe punt with detachable inboard motor and a patented lightweight collapsible boat, dubbed the 'Goatley', a small wood and canvas assault craft carried by infantry units enabling them to cross waterways as they pursued enemy forces who, in their retreat, had destroyed bridges and other means of crossing rivers and canals. Weighing around 330lb and measuring 15.8ft by 4.4ft by 2ft, each boat could carry ten fully equipped troops. The design of the craft resulted in orders for hundreds of examples from Saunders-Roe and even more when Fred Goatley later introduced an improved and strengthened version.

Goatleys have featured in films, notably *A Bridge Too Far* from 1977, directed by Richard Attenborough with an all-star cast including Sean Connery, Michael Caine, Lawrence Olivier, Gene Hackman, Robert Redford, Anthony Hopkins, Dirk Bogarde and many others.

Fred Goatley's greatest achievement was the collapsible canoes he designed in the ensuing war for commando units and elite special forces groups operating covertly behind enemy lines and within enemy-held ports and harbours. Inspired by Major Herbert George Hasler of the Royal Marines, three different unique, flat-bottomed craft were designed by Fred Goatley, two of which were adopted by the War Office, the Mk II 15ft canoe and the

Mk II★★ 17.3ft three-man version, both of which were designed to collapse vertically for easy stowage aboard submarines but which could be erected in just 30 seconds. Over 1,100 were built at Saro Laminated Wood Products and elsewhere. Six of the Mk II★★ craft, the now world-famous Cockleshell canoe, were employed in the equally notorious Operation Frankton, the assault on enemy cargo shipping in Bordeaux docks, on the River Gironde, in December 1942. The raid was subsequently recognised in the 1955 movie *Cockleshell Heroes* starring José Ferrer and Trevor Howard.

Besides the Bordeaux raid, Cockleshell canoes were used for other clandestine operations in harbours and rivers around the world and they were also deployed to assess the D-Day landing beaches prior to the launching of Operation Overlord. Arising from his work on commando canoes, Fred Goatley came to be affectionately referred to as 'Father Cockle'.

Apart from the canoes and collapsible assault boats, Fred Goatley designed more than ten other types of boat for military applications including a patented unsinkable lifeboat which could self-seal bullet holes, a collapsible pontoon, a 25ft rescue powerboat and a surf landing craft.

In summarising the many achievements of this unsung local hero, author Quentin Rees has said, 'What he did for the war effort was enormous [but] even to this day he is unrecognised for his efforts, which were substantial.' A plaque erected near the Folly Inn at Whippingham is a small but important token of recognition for this talented man.

Where the Cockleshell canoe was a surface craft, exposing its occupants to the risk of observation and interception, a submersible craft for special operations was seen as being a desirable alternative, allowing underwater reconnaissance or attacks with limpet mines to be carried out from beneath the surface. Conceived by Major Hugh Quentin Reeves, the aluminium, electrically powered motorised submersible canoe (MSC), or 'Sleeping Beauty' as they were dubbed, was placed in production at Saunders-Roe's East Cowes works for the Special Operations Executive from around 1943, a total of fifteen being completed. Craft of this type, capable of submerging to a depth of 40ft, piloted by a commando in frogman's outfit, were deployed for Operation Rimau, the raid on Japanese shipping in Singapore in September 1944.

Post-war initiatives with specialist marine vessels included an unusual, experimental craft for the Naval Intelligence Department at Keyhaven, designed and constructed by Saunders-Roe. Known as the WALRUS (Water and Land Reconnaisance Unit Survey), it was a small amphibious vehicle powered by a Coventry Climax petrol engine which drove two outboard contra-rotating, tapering helices which ran the length of each side of the 18ft-long hull.

A major success story in respect of a distinctive, high-quality working boat design owes its inspiration to Sir Peter Thornycroft. It was named the Nelson workboat after one of the two foremen who helped set up the Keith, Nelson & Company boatyard at Bembridge where the craft were built: Keith Butt and Arthur Nelson Compton.

These instantly recognisable boats with varnished timber and painted GRP hulls have been procured by discerning marine operators worldwide and have seen service with customs, naval and harbour authorities throughout the UK and overseas. Trinity House has used them extensively as pilot tenders and the police have employed them in various coastal locations such as the Solent and the River Thames for patrol and surveillance work. The original designs, the Nelson 32, 34, 40 and 45, were the work of Peter Thornycroft. Later designs were produced by John Askham, Peter Collet, Rolland Paxton and Arthur Mursell. Literally hundreds of these boats have been built, a prodigious output, and they are still being built, now at the boatyards of Seaward Marine in West Cowes and Dale Nelson in Pembrokeshire, as well as in the Netherlands and in Guernsey. Though Nelson boats are no longer built at Bembridge, the associate business, T.T. Boat Designs, continues to operate there.

This silver model of the standard 30ft Lamb & White lifeboat was presented by John White to Andrew Lamb, Southampton, in appreciation of his services and 'To record the designing of the above boat and its introduction into the P&O Company's fleet in 1846'. Supplied for shipboard use as well as for shore rescue stations, it was not intended to be self-righting but was virtually non-capsizeable. (Isle of Wight Heritage Services)

For its time, the Lamb & White lifeboat was probably the most commonly built type of all pulling lifeboats at well over 800. The Admiralty alone ordered 500 for Royal Navy ships. Produced in several variants, this drawing by Forbes & Bennett, which comes from John White's original patent registered in 1846, shows the clench built lifeboat gig version as supplied to the Royal Yacht Squadron. Note the buoyancy cells under the gunwale. (Isle of Wight Heritage Services)

The Classic Boat Museum at Cowes has this example of a post-war airborne lifeboat Mk III on display. Conceived by Uffa Fox, the airborne lifeboat saved the lives of many aircrews during the war. The Mk III version, which measured 30ft in length and 7ft beam and fully equipped had an all-up weight of 4,460lb, and was made from aluminium. The original airborne lifeboats had wooden construction. (David L. Williams)

Constructed in quantities in the 1950s by Saunders-Roe at Beaumaris, the Mk III variant of the lifeboat was produced for Avro Shackleton coastal reconnaissance aircraft. The lifeboat had a capacity for a maximum of ten persons. They carried oars, a mast and sail, plus a Vincent HRD 155hp engine. Compartments in the hull structure contained provisions and other essential survival equipment. (Crown Copyright)

The collapsible Cockleshell Mk II canoe, designed by Fred Goatley, was the craft utilised in the Operation Frankton raid on Bordeaux in December 1942. Here, one is seen under construction probably at the Saro Laminated Wood Products plant at Whippingham, its canvas sides evident. (Combined Military Services Museum)

A completed Cockleshell Mk II canoe photographed alongside the River Medina at Whippingham. One of these craft is preserved at the Combined Military Services Museum at Maldon, Essex. (Combined Military Services Museum)

A 'Sleeping Beauty' one-man motorised submersible craft (MSC) shown submerged and descending. The pilot in his frogman's suit can be seen at the controls. Despite wearing goggles, visibility at depth and in murky water would have been a challenge. Fifteen of these craft were built by Saunders-Roe. (Australian War Memorial)

A wartime photograph showing a Goatley with six soldiers aboard during a dry-land demonstration to an infantry section from 'B' Company, 1/6th Queens Regiment, in preparation for an impending assault on enemy lines. The demonstration appears to be causing considerable amusement. In practice, despite its lightweight construction, the Goatley needed at least this many men to carry it any distance. Fred Goatley's design for the collapsible boat, at one time erroneously designated the Cockleshell Mk IV canoe, was patented in March 1938 (No. 480987). (Crown Copyright)

Here Belgian para-commandos are seen with their Goatley boats during a river crossing exercise on the River Meuse, Belgium, in 1957. One of these Goatleys is preserved at the Commando Museum at Flawinnes, near Namur, Belgium. (Commando Museum, Belgium)

A classic Nelson workboat, the pilot boat *Gazelle*, completed for the Forth Pilotage Authority, during pre-acceptance trials. (Author's collection)

113

Above: Another example of the celebrated Nelson design, the Trinity House pilot cutter *Vagrant*. (Chris Bancroft)

Opposite top: John Fairfax tries out his 25ft 860lb rowing boat *Britannia* near the Clare Lallow boatyard at West Cowes on 23 December 1968, with designer Uffa Fox (centre) looking on from the launch. John Fairfax's celebrated solo crossing of the Atlantic took from January to July of the following year. *Britannia* is now an exhibit at the Classic Boat Museum. (Author's collection)

Opposite bottom: *Britannia II* rowed by John Fairfax and Sylvia Cook off the California coast in April 1971 at the start of their epic 8,000-mile Pacific crossing which took 361 days to complete. They reached Hayman Island, Australia, on 22 April 1972. Like the earlier *Britannia* this craft was also designed by Uffa Fox and built by Clare Lallow. (Author's collection)

Left: One of the first Island-built lifeboats constructed for the RNLI, the seventh to be completed by Hansen at Cowes, in 1888, this is the standard 39ft twelve-oared pulling self-righter *George Moore II* based at the Porthdinllaen station in Wales. Launched only eleven times, it saved twenty-two lives. (Tony Jones, Rhiw website)

Left: The steam lifeboat *James Stevens No. 3* was one of four similar boats constructed from 1898 by J. Samuel White which included the first three craft completed by the company for the RNLI. (RNLI Angle Lifeboat Station)

Opposite top: Just over fifty years later, J. Samuel White celebrated the delivery of its 100th RNLI lifeboat, the Barnett (Beach)-class *Hilton Briggs*, placed on the Aberdeen station. (Isle of Wight Heritage Services)

Opposite bottom: The Watson-class lifeboat *TGB*, built by White's in 1962, is now preserved at the Scottish Maritime Museum at Dumbarton. While based at Longhope station, the *TGB* overturned during an attempted rescue in the most appalling weather conditions on 17 March 1969, a tragedy which cost the lives of her entire crew. The disaster acted as a catalyst for the development of a self-righting modification for Watson-class non-self-righting lifeboats. (Alan Kempster)

R.N.L.B. "Hilton Briggs"
The 100th Lifeboat built to the order of the Royal National Lifeboat Institution, for the Aberdeen Station.

SAMUEL WHITE

J. SAMUEL WHITE & CO., LTD., COWES, ISLE OF WIGHT, Shipbuilders and Engineers.
London Office: 8 Duncannon Street, W.C.2

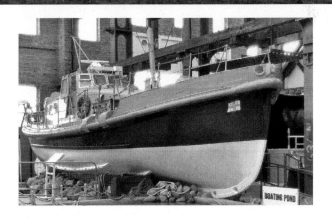

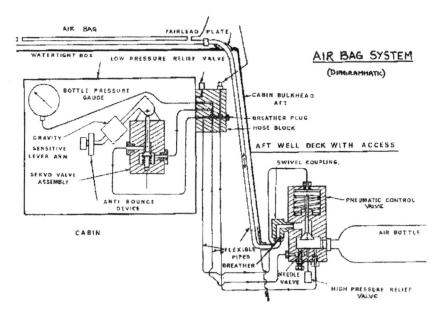

Devised and developed by engineers at the British Hovercraft Corporation's Test Facilities Division, formerly EEL, this practical self-righting system comprised an air bag atop the cabin which would be inflated from gas bottles activated by a gravity valve if a boat rolled more than 120 degrees from upright. (Stuart Welford, courtesy Royal Institution of Naval Architects)

One hundred and fifteen years after the first RNLI lifeboat was built at Cowes, the last of 419 boats to enter the water is believed to have been the Yarmouth station's large 55.8ft Severn-class boat *Eric and Susan Hiscock*, completed by Souters in 2002. (Colin Wilkinson)

6

JETS, ROCKETS & THE RACE INTO SPACE

Apart from expediting landplane development, as already mentioned, the Second World War also fostered rapid advances in rocket and jet technology, areas of engineering expertise that continued with rapid progress in the immediate post-war years. While the USA and USSR may have had the highest profile and most expensive programmes involving this technology, driven by Cold War competition, the United Kingdom also retained a strong presence having led the way with the invention of the jet engine. In the 1950s, on two occasions, British aircraft captured the world air speed record, culminating with the record speed of 1,132mph set by the supersonic Fairey *Delta* on 10 March 1956.

With the gradual decline of its flying boat business, Saunders-Roe became engaged first in jet aircraft production and then, later, in space rocket activity. Both diversions were rather surprising for a business which was located by the sea without either a readily accessible runway of adequate length or a launch site from which rockets could be fired. This did not hamper Saunders-Roe in the slightest, though. On completion, its jet aircraft were transferred to Boscombe Down in Wiltshire for flight acceptance trials and the rockets underwent static engine-proving tests at High Down, at the western extremity of the Isle of Wight, prior to shipment to Australia where they were launched from the Woomera rocket site.

Before that, going back to the end of the Second World War, when the conflict with Japan was still raging, a need was identified for a fast and powerful fighter aircraft, ideally jet-engined, that could operate independently of airstrips or aircraft carriers, in the former case because many had been destroyed by the retreating Japanese as Pacific islands were abandoned, and in the latter because the carriers that were available were not equipped for such aircraft and were already committed to fleet action. Saunders–Roe responded to Air Ministry specification E6/44 with its proposal for a flying boat fighter, the SR–A/1, later dubbed the 'Squirt'. Development time for such a project meant, however, that physical aircraft, three of which were built, were not ready for service until after the war had ended.

This change of circumstances eliminated their primary *raison d'être* and as highly specialised aircraft they could not readily be adapted for a different role. Thus, they became one of a series of projects that were either cancelled or rendered redundant in this period. Nonetheless, despite suffering this setback, the experience gained with the SR–A/1 placed Saunders–Roe in good stead for the jet aircraft designs that followed.

Simultaneously, plans were unfolding for streamlined jet-engined commercial flying boat transports, notably the elegant medium-range Duchess and the incredible P.192 (unofficially given the name 'Queen'), the largest practical aircraft ever conceived.

Inspired by the expressed interest of the P&O Line for an aircraft suitable for the Dominion run, at that time dependent on ocean-liner traffic, the latter was to be capable of carrying a so-called 'liner load', that is 1,000 passengers on the run from Southampton to Sydney in a maximum of 48 hours. The study that was undertaken resulted in the extraordinary P.192 jet-powered design which, though only a concept, would have dwarfed anything then flying and which, still today, remains bigger than any aircraft that has so far been built – see the table that follows.

An aircraft of the size of the P.192 could not have been built without significant enlargement of the Columbine hangar, an endeavour that might have been constrained by the lack of available space on the East Cowes foreshore. In the event, despite its practical and economic feasibility, the P.192 never materialised because it probably proved to be too radical, even for the forward-looking P&O Company.

Yet more phenomenal was a large nuclear-powered flying boat conceived by Henry Knowler with banks of engines mounted in the tailplane. At that time, the possible applications of atomic energy seemed limitless but while this concept may have been feasible, offering continuous operation without the need to refuel, it did not appear to take account of the implications of radioactive contamination that would have resulted in the event of an accident.

In a sense, the demise of these great and fantastic but unfulfilled schemes signified the bringing down of the curtain on the Saunders–Roe flying boat era.

The first of the company's supersonic machines to come off the drawing board was the striking SR.53 mixed-power (combined jet and rocket propulsion) interceptor fighter which first flew in May 1957, a sleek but powerful aircraft capable of a speed of Mach 2.0 (1,520mph) and able to climb to 50,000ft in just over 2 minutes. Although the second of the two prototypes that were built experienced a fatal crash, it was a proven design and it was described by the test pilot as 'extremely docile and exceedingly pleasant to fly with very well harmonised controls within the flight envelope'.

Irrespective of that, it was considered to be too light and underpowered for the envisaged equipment fit and production did not proceed beyond the prototypes. Fortuitously, the Saunders-Roe design team, led by Maurice Brennan, had been working in parallel on a developed machine, the larger and faster SR.177. This had a design maximum speed of Mach 2.35 (1,785mph) and the capability of climbing to 60,000ft in just 3.1 minutes. Orders were received for an initial nine production aircraft for the RAF with an option for a further twenty-seven. The German Air Force also registered a positive interest in taking the SR.177 while a naval version for operation on the Royal Navy's two large carriers HMS *Ark Royal* and *Eagle* was also in the planning stage.

All was set fair for this groundbreaking aircraft design from Saunders-Roe when, to use another metaphor, it was struck by a bombshell. An ill-conceived and ill-timed government white paper published in April 1957, a review of defence requirements commissioned by Duncan Sandys, the Minister of Defence, essentially destroyed the project. It declared that all fighter aircraft were obsolete, as they would, in due course, be replaced completely by ground-to-air missiles. It was hoped that the five airframes already in an advanced state of construction could be completed and delivered to the RAF and that the follow-on aircraft, for which materials had been procured, would see service with the German military but, on 23 December 1957, the company was advised that there would not be a German order and, on the following day, that the already truncated RAF contract was cancelled completely.

The incomplete SR.177 aircraft were dismantled, new production and treatment techniques were abandoned and 1,470 workers, many of them highly skilled, lost their jobs, all on a misguided and inaccurate evaluation of future defence requirements which was subsequently reversed, though by then, for Saunders-Roe, it was too late.

Fortunately, arising from the knowledge it had gained from working with rocket fuels for these mixed-fuel aircraft, Saunders-Roe had been working in parallel with the Royal Aircraft Establishment, Farnborough, on other projects. This had led to an invitation in 1955 for the company to build an upper-atmosphere research rocket, the Black Knight, initially as a test bed for the medium range ballistic missile Blue Streak, concurrently under joint development by de Havilland and Hawker Siddeley Dynamics.

Twenty-five Black Knight rockets were built, twenty-two of which successfully launched, the first in September 1958, and after the Blue Streak was abandoned as a nuclear payload missile, the Black Knight was considered, along with the Blue Streak, for a combined satellite launcher, code-named Black Prince. In this vehicle, the Black Knight would have formed the second-stage propulsion unit. The concept, which later became known by the project name Black Prince, had been first mooted in 1957. It was revived in April 1960 as a proposed joint development by the Royal Aircraft Establishment, Saunders-Roe and Bristol Siddeley after the Blue Streak missile had been cancelled.

Although the Black Prince project was also shelved with Blue Streak being used instead as the first stage of the European Launcher Development Organisation (ELDO) rocket Europa, the case for an independent British satellite launch vehicle had, it seemed, been made. Consequently, arising from its successful track record with the Black Knight, the company, by then trading as British Hovercraft Corporation, was awarded the contract to build five three-stage Black Arrow rockets.

This programme culminated in the launching of Black Arrow R.3 on 28 October 1971 placing the Marconi-manufactured Prospero satellite into orbit, the first and only all-British space achievement. Sadly, the euphoria at this immense achievement was restrained to put it mildly given that the government of the day, as on previous occasions, had intervened to terminate the project three months earlier on 29 July 1971 on the basis that the UK had no need for an independent space research programme. Diplomatically, the company accepted the government's decision with little public comment, leaving others to express regret concerning its bittersweet achievement and condemnation of the official shortsightedness.

One commentator later summed it up with these words: 'The UK remains the only country to have successfully developed and then abandoned a satellite launch capability.' Another said, 'Inexplicably, despite foreseeing the amount of demand there would be for satellite technology, considering the imminent explosion in demand for international telecommunications satellites, the government decided to stop all research into this extremely profitable business.'

Through its experience in space engineering and rocketry, the company did secure the contract in November 1972 to design, produce and launch the AP.135 Falstaff space systems rocket for the Ministry of Defence. Twelve of these rockets were produced but the first six firings were so successful that the final six were never launched. But hope for continued participation in future rocket and space equipment development was dashed by more government negativity.

Later, in August 1987, a decision against participation in an expanded European space programme, through the government's refusal to pay contributions, was described as 'astonishing'. It denied British companies, and notably Westland Aerospace, as Saunders-Roe was by then called, from involvement in its European Space Agency's *Hermes*, *Columbus* and *Ariane 5* projects.

The damage inflicted on industry, quite apart from the wasted money, by repeated governmental cancellations of high-profile cutting-edge research and development programmes requires little imagination to appreciate – not only the loss to the country as a whole of critically valuable technological expertise in markets that have since blossomed to become major revenue earners but also the huge levels of redundancy among the local workforce and the consequent depression to the morale of remaining employees. The experience of Saunders-Roe and British Hovercraft Corporation during the period from the 1950s to the mid-1970s sowed a seed of cynicism amongst local workers with regard to the durability of government contracts. It should be noted that the UK's space programme, essentially Saunders-Roe and BHC's, was the most efficient in the world with an over 92 per cent success rate in rocket launches, thirty out of a total of thirty-three.

Before moving on to look at the emergence of hovercraft as the next chapter in the Saunders-Roe story, it is fitting here to relate something of the background of the man who was at the helm throughout the period of the jet and rocket projects, responsible for the design work on the SR.53, SR.177 and Black Knight. This was Maurice Brennan, chief designer at Saunders-Roe from October 1953 to March 1959.

Born in London in April 1913, Maurice Brennan was educated at St Mungo's Academy, Glasgow, and Glasgow University. After graduating, he joined Hawker Aircraft in 1934 remaining there until 1936 when he transferred to Saunders-Roe as one of Henry Knowler's design team. He worked on the S.36 Lerwick and Short Shetland flying boats, serving in the Stress, Aerodynamics and Projects offices. From 1947, Maurice Brennan was appointed technical assistant to Sir Arthur Gouge, then the vice chairman and chief executive, taking charge of helicopter development. Promoted to chief designer, he took responsibility for the interceptor fighter projects and the Black Knight programme. He also led the detail design work on the SR.N1 hovercraft until his resignation in March 1959 when he left to join Vickers Armstrongs.

Although it does not readily fit into the chapter structure of this book and although the production and much of the design work took place at Eastleigh in Hampshire, Saunders-Roe's activity in respect of helicopter development should not be overlooked. After taking over the Cierva company in January 1951, Saunders-Roe took over all design work on the W.14 Skeeter helicopter under project number P.502. This resulted in the production of eighty-eight machines for operation by the RAF, the Army and the Federal German military in pilot training and air observation roles. The Skeeter was followed by the larger P.531 Sprite produced for military and civilian purposes. Some six aircraft had been completed by July 1959 when, following the takeover of Saunders-Roe by Westland Aircraft Limited, all subsequent development was transferred to the new parent company in the form of the Westland Wasp and Westland Scout with the design authority moved to Yeovil. Nonetheless, the credit for the success of this helicopter, of which around 280 were ultimately built in the two variants, firmly belongs with Saunders-Roe.

Of interest, as part of Saunders-Roe's helicopter activities, the company also developed two pulse jet-boost engines, designated P.J.1 and P.J.2, which were trialled on two Mk VI Skeeters G-AMTZ and G-ANMI, mounted on the tips of their three-blade main rotors.

In the later stages of the Second World War, a need arose for a powerful fighter/reconnaissance aircraft that could operate from remote islands in the western Pacific as part of the final phase of the battle against the Japanese. To meet this need, Henry Knowler's team designed and built the Saunders-Roe SR.A/1 'Squirt' jet-powered flying boat, the only aircraft of its type. Here, the third of the three aircraft built, TG271, commences her take-off run. It crashed and sank in the Solent while landing on 12 August 1949. (Saunders-Roe)

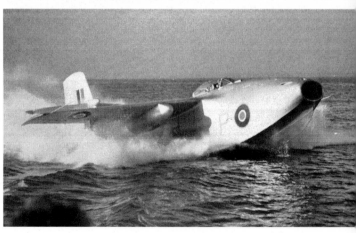

Above: An impression of the elegant medium-range Duchess civil flying boat transport (project No. P.131) announced in the June 1950 edition of the Saunders-Roe magazine *The New Slipway*. The 58-ton *Duchess* would have had six Rolls-Royce Avon jet engines giving a cruising speed of 500mph. The wingspan would have been 135ft, the overall length 124.5ft and within the accommodation cabin seating for seventy-four passengers. (Saunders-Roe)

Opposite top: Looking, perhaps, slightly cumbersome on the water, the SR.A/1 made a graceful and striking impression when in flight. This is the first machine, TG263. The SR.A/1 was 50ft long, 46ft in wingspan and was powered by twin Metrovick Beryl axial-flow turbojet engines producing 3,850lb thrust and providing a maximum speed of 512mph. (Saunders-Roe)

Opposite bottom: A contrast in sizes. The SR.A/1 looks diminutive alongside the giant Princess at barely one-twentieth of the bigger craft's size as they stand alongside each other inside the entrance to the Columbine hangar. (Saunders-Roe)

Above: Despite the abandonment of the Princess, Saunders-Roe pursued an even more ambitious flying boat project for what would have been the largest civil aircraft ever contemplated, by a significant margin. Invited by the P&O Line, Saunders-Roe performed a feasibility study and preliminary design for a plane for the route to Australia which, at that time, was a challenge for contemporary landplanes and involved a six-week voyage by ocean liner. Referred to unofficially as the Queen, this giant aircraft reveals how, long before the event, Saunders-Roe had perceived the future scale of mass transportation by air. This impression, as well as the next illustration of the unequalled P.192, comes from the preliminary feasibility report TP.166 'P.192 Passenger-Cargo Airliner', dating from 1956. (Saunders-Roe)

Opposite top: Capable of completing the 10,500-mile run in 48 hours with stops en route, the P.192 Queen had capacity for up to 1,100 passengers on four decks – a 'liner load'. The engines would have been twenty-four Rolls-Royce Conway bypass turbo-jet engines, buried in the wings. The accommodation was for first- and tourist-passengers, in comparable ratios to those of a ship. No less than twenty-seven toilets would have been provided, essential for such a large complement! (Saunders-Roe)

Opposite bottom: The P.192 had staggering dimensions. To convey an impression of its huge scale, this table compares it with other large aircraft, including the Airbus A380, currently the world's largest airliner, and Howard Hughes' 'Spruce Goose' flying boat. While its cruising speed would have been 30 per cent slower than either the 747 Jumbo or Airbus A380, its passenger payload would have been over 30 per cent greater.

Acquired in 1988, the current owner of the iconic *Bluebird* K3 is the Foulkes-Halbard Collection at Filching Manor, East Sussex. Their dedicated work in resurrecting her and restoring her to operational condition took twenty-two years to complete. This view is from October 2012. (Lisa Beaney Photography)

An advert for the Princess flying boat from 11 June 1948. The impression depicts the giant aircraft aloft over New York's skyscrapers as a luxury passenger transport operating the prestigious transatlantic route which had been maintained in the years prior to the Second World War by the famous Boeing Clippers. (Aviation Ancestry)

Right: The launch of Black Arrow R3 from Woomera, Australia, on 28 October 1971, carrying the satellite *Prospero* into orbit. It was Britain's greatest and only all-domestic space achievement. (British Hovercraft Corporation)

Below: Saunders-Roe's hopes were high in the late 1940s for both the Princess flying boat and the unique SR.A/1 flying boat jet fighter. This advert, dated 3 September 1948, promotes both the latter aircraft and its engines supplied by Metropolitan Vickers. (Aviation Ancestry)

Opposite top: A specially modified Britten-Norman Islander, registration G-BVHX, designated BN-2T-4R, was fitted with a purpose-built Westinghouse Electronics Systems' Multi-Sensor Surveillance Aircraft system (MSSA). Modifications involved strengthening of the fuselage, incorporation of a modified Trislander main wing and the installation of twin uprated Rolls-Royce Allison 250-B17F turbo-prop engines. The first of the type made its maiden flight on 29 July 1994. (Britten-Norman)

Opposite bottom: The Britten-Norman BN-2T-4S Defender 4000 prototype G-SURV flying past Freshwater, Isle of Wight. By the year 2000, well over 10 million flying hours had been logged by BN-2 series aircraft. (Britten-Norman)

ARV Super-2 G-BPMX flies past during the Great Vintage Flying Weekend at Kemble Airport, Gloucestershire, in May 2009. These aircraft were 18ft in length, had a wingspan of 28.5ft and were powered by a Hewland AE75 3-cylinder liquid-cooled inline engine rated at 77hp. Production had hardly begun when the manufacturing rights were acquired by Opus Aircraft, an American company based in North Carolina, ending the project on the Island. (Adrian Pingstone)

The SR.N4 Mk 3 or Super-Four *Princess Anne* powers along the English Channel returning to Dover following her conversion from Mk 1 standard. Over their first twenty-five years of service, the SR.N4 fleet carried 40 million passengers across the Channel. (British Hovercraft Corporation)

The API-88-100 hovercraft *Freja Viking* of Scandinavian Airways Systems skimming over ice near Kastrup Airport, Copenhagen. (Westland Aerospace)

The powerboat *Avenger Too*, designed by Don Shead and built by Souters of Cowes, won the Round Britain Race run between July and August 1969. With propulsion supplied by three 125hp Mercury outboard engines and driven by Timo Makkinen, Pascoe Watson and Brian Hendicott, she completed the 1,459-mile circuit of the country in just over thirty-nine hours at an average speed of 37.1 knots (42.6 mph). (Graham Stevens)

The Project Thrust team relocated *Thrust 2* to the Black Rock Desert in Nevada where the world land speed record was finally secured on 4 October 1983. *Thrust 2* is now on display in the Coventry Transport Museum. (John Ackroyd)

The incomparable classic offshore powerboat *Surfury* seen at speed. Her extraordinary, low-resistance, streamlined hull form gave her the edge over more powerful contemporary challengers. (Graham Stevens)

The Sea Cadets Association's brigantine training ship TS *Royalist*, designed by Colin Mudie and launched by Princess Anne at Groves & Guttridge, East Cowes, on 3 August 1971, was the first new square-rigged sail training ship to be built in the United Kingdom since before the Second World War. The *Royalist* has provided seagoing experience for more than 30,000 Sea Cadets over her forty-three-year career, which ended with her decommissioning in November 2014. She measured 83 gross tons with main dimensions of 97ft overall length and 76.5ft hull length. (Graham Stevens)

The BHT 130 hovercraft *Solent Express*, completed in 2009, is the latest passenger-carrying hovercraft to be built on the Isle of Wight. Her welded hull was constructed by Aluminium Shipbuilders at Fishbourne with fitting-out and engine installation performed by Hoverwork at St Helens. Derived from the earlier API-88 craft, the BHT 130 has numerous enhancements. It has greater capacity, improved all-round cockpit visibility, more powerful engines and a stylishly remodelled and streamlined exterior. At 70 tons all-up weight and 96.4ft length, her four MTU turbo-charged diesel lift and propulsion engines, rated at 4,250bhp, give a service speed of 45 knots (51.7 mph). (Griffon Hoverwork)

Another impression of the planned Saunders-Roe jet-engined medium- or long-range *Duchess* flying boat, featured on the front cover of the *The Aeroplane* magazine of 27 July 1951. (Author's collection)

The *Vestas Sailrocket 2* photographed at speed on 24 November 2012 while breaking the world outright sailing speed record for 500m at 65.45 knots (75.2 mph). (Helena Darvelid, Vestas Sailrocket)

An Enfield 8000 electric city car (ECC) breaking the European speed record for a battery-operated vehicle! Incredibly, that is exactly what happened on 5 September 2015 when an original 1970s Enfield car, modified by motoring journalist Jonny Smith into the Flux Capacitor, achieved 118.38mph on a ¼-mile dash at the Santa Pod racetrack, in Wellingborough, Northamptonshire. As such, it is now the fastest legal electric vehicle in Europe. (Mark Skinner)

An SR.N6 hovercraft of Hovertravel at the company's Ryde, Isle of Wight, terminal. Prior to the introduction of the diesel-powered API-88, these gas-turbine engine craft were operated on the Ryde to Southsea route. They were also used by British Rail Seaspeed for a regular service between Cowes and Southampton. (Author's collection)

The Princess flying boat G-ALUN making a low-level pass during the display at the Farnborough International Air Show of September 1952. (Author's collection)

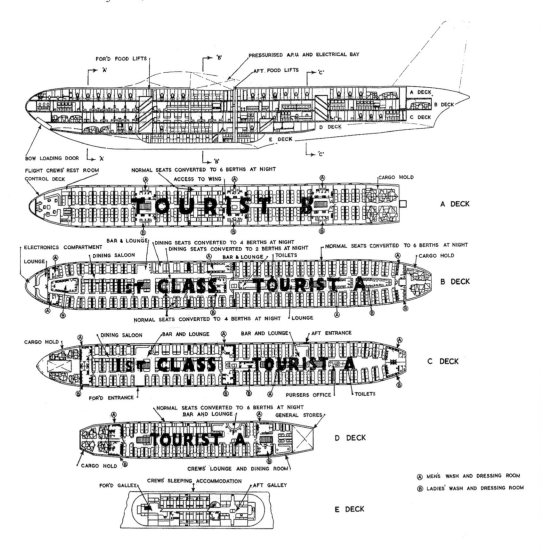

	All–Up Weight	Length	Wingspan	Passengers	Max Speed
P.192 'Queen'	670 tons	318ft	313ft	1,100	448mph*
Hughes H4**	177 tons	218ft	320ft	750 troops	250mph*
Boeing 747-8	440 tons	250ft	225ft	605	614mph
Airbus A380–800	573 tons	239ft	262ft	853	634mph
Antonov Mriya	640 tons	275ft	290ft	cargo	528mph

* cruising speed ** the H4 flew only once on 2 November 1947 for 1 mile at 70ft

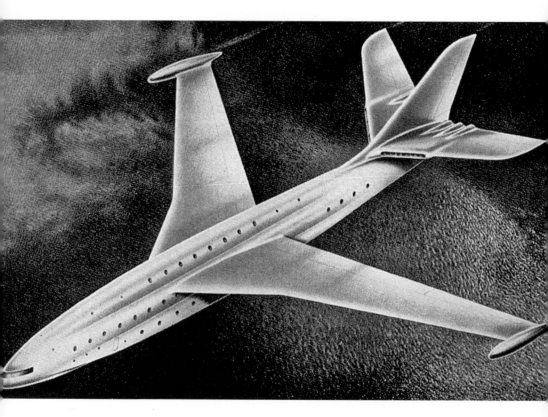

Above: Henry Knowler's concept for the 'airliner of the future' was this atomic-powered flying boat with sixteen engines mounted in the tail wing. It was projected around the time when the Princess flying boats were also under consideration for adaptation with nuclear engines. Although no dimensions are quoted for the concept, with three or more passenger decks it would have been a giant aircraft. However, the days of the flying boat were over. (Saunders-Roe)

Opposite top: With the end of the flying boat-era Saunders-Roe next turned to jet fighters and rocket vehicles one of the results of which was the delta-wing, mixed-power – jet and rocket – interceptor fighter SR.53. Here, manufacture of the SR.53 is under way in a partitioned assembly area in the Columbine hangar at East Cowes. Presumably the screening was a security requirement stipulated by the Air Ministry. (Saunders-Roe)

Opposite bottom: The first SR.53, registration XD145, is seen on the tarmac at Boscombe Down, Wiltshire, from where flight tests took place. Measuring 45ft in length and 25ft wingspan, the SR.53 had an Armstrong Siddeley Viper turbo-jet delivering 1,640lb thrust and a de Havilland Spectre rocket motor delivering 8,000lb thrust. Only two of these aircraft were built, the third machine cancelled when effort was switched to the larger SR.177. (Saunders-Roe)

Above: The nearest thing to a view of a completed SR.177 aircraft were the full-scale mock-ups: two of them shown here in the Folly works at Whippingham. The intended power plant of the SR.177 was a de Havilland Gyron turbo-jet delivering 14,000lb thrust and an up-rated 10,000lb thrust de Havilland Spectre rocket motor. (Saunders-Roe)

Opposite top: An impressive sight in flight, the SR.53 was a truly stylish aircraft. The single surviving SR.53 mixed-power interceptor can now be seen in the collection of the RAF Museum at Cosford, near Wolverhampton. (Saunders-Roe)

Opposite bottom: The SR.53 interceptor fighter was in effect the prototype machine for a planned production version, the SR.177 which never materialised, falling the victim of a muddled defence review. This impression of the SR.177 aircraft shows it in the livery of the German Air Force, one of the anticipated customers. The SR.177 was 5ft longer and 2ft wider across the wingspan than the SR.53. Armament for these aircraft was wing tip-mounted Red Top guided missiles. (Saunders-Roe, courtesy of Ray Wheeler)

Space activity began in the form of the 33ft Black Knight ballistic research rocket of which twenty-five were completed. The Black Knight shown here is at Woomera prior to launch. The power plant was a 16,000lb-thrust Gamma II rocket motor burning high test peroxide (HTP) and kerosene. The Black Knight could reach an altitude of 70 miles. (Saunders-Roe)

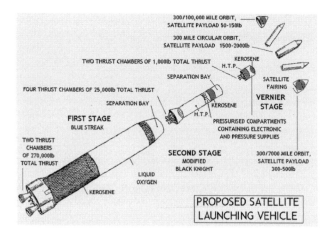

Exploded view of the planned Black Prince satellite launcher showing the component parts, including the modified Black Knight second stage and various satellite payloads. It is quite possible that, had the Black Prince gone ahead, final assembly would have taken place on the Island with static testing conducted at the established High Down rocket test site. (Author's collection)

The positive Black Knight programme led on to a government contract for a much larger, low-cost satellite launch vehicle, the 43ft three-stage Black Arrow. The photograph shows the detail of the 50,000lb thrust Rolls-Royce Gamma 8 first-stage propulsion unit of the never-launched Black Arrow R4 which is now on display at the Science Museum, London. (Author's collection)

Above: A Skeeter helicopter, developed by Saunders-Roe from the Cierva W.14 design. It was followed by the larger P.531 Sprite. Both were successful machines. (Author's collection)

Opposite top: All rockets were statically tested at High Down, above Alum Bay prior to shipment to Australia where they were launched into space from the Woomera launch site. This culminated in the successful launch of the fourth Black Arrow, R3, on 28 October 1971, placing the Marconi-constructed *Prospero* satellite into earth orbit – the only ever all-British space achievement. The Black Arrow second stage comprised a 15,340lb Rolls-Royce Gamma 2 motor and the third stage was an RPE Waxwing solid propellant motor delivering 6,130lb thrust. Note the Needles Lighthouse and rocks in the distance to the left of this view of a Black Arrow static test. (British Hovercraft Corporation)

Opposite bottom: One of the Falstaff research rocket launches which followed the Black Arrow cancellation. These 24ft-long rockets, twelve of which were built but only six fired, had a distinctive bulbous nose which had clam-shell fairings enclosing the payload bay from which experimental payloads were released. Thrust was provided by a Stonechat solid fuel rocket motor. Note the pronounced cruciform fins. (British Hovercraft Corporation)

7

SKIMMING THE SURFACE

Following the loss of the SR.177 aircraft contract which had caused 1,470 redundancies, equivalent to a substantial percentage of the Island's total industrial workforce of today, Saunders-Roe was selected in the same year to build and test three models of a new type of waterborne craft that had been invented and patented by the enterprising engineer Sir Christopher Cockerell, derived from the experiments into air cushion lift that he had been conducting since 1955. It was an order that came to the company's rescue at a difficult time and which, ultimately, would restore the company's fortunes and eventually create many new jobs.

In this new type of craft an annular jet of air provided both lift and lubrication while an air propeller provided thrust. Effectively, having no drag whatsoever, such a craft was capable of much higher speeds for a given power output than an equivalent displacement craft.

Christopher Cockerell was born in Cambridge in 1910 and educated at Gresham's School, Norfolk, and later at Cambridge University where he studied mechanical engineering, radio and electronics. Subsequently, he worked for Marconi in its research centre at Writtle, near Chelmsford, Essex, from 1935 to 1951, engaged in the development of radar and radiolocation technology. After he left Marconi, inspired by some of the studies into marine craft motion that had been carried out by John I. Thornycroft, he began to investigate alternative means by which the speed of marine craft could be increased. Ultimately, through practical experimentation, this work led to his patented invention (No. 854211) of air cushion technology and the first practical hovering craft.

The potential of the hovercraft was enormous and after demonstrations of the concept were performed for the Admiralty and Air Ministry, the all-important initial research and development contract was awarded to Saunders-Roe by the Ministry of Supply. Saunders-Roe had actively promoted itself for this project but its selection recognised the company's innovative culture working with groundbreaking concepts and cutting-edge technology.

Subsequently, in October 1958, the National Research Development Council (NRDC) through its subsidiary, Hovercraft Development Limited (HDL), placed a contract with Saunders-Roe for a prototype experimental, manned craft, the Saunders-Roe Nautical 1 or SR.N1, the world's first hovercraft. The craft's designation reflected the fact that Cockerell's basic concept had required the development of a totally original design by the Saunders-Roe design team to produce this first practical, air cushion hovering craft. Thus Saunders-Roe and the Isle of Wight became the birthplace of the hovercraft.

The SR.N1, a fully amphibious craft, employed inward-directed peripheral air-jets to raise it above the surface of the water. As progressively developed, flexible rubber skirts – another Saunders-Roe invention – were attached to the hull to help contain the air cushion and these featured on all subsequent hovercraft built by the company.

Recognising the inherent potential of this new form of water transport, the company immediately embarked on ambitious plans to apply the principle to much larger vehicles for both military and civilian roles. The rapidity of hovercraft development was astonishing by any measure. In just three years from the first 'flight' of the SR.N1 on 11 June 1959, a fare-paying commercial passenger hovercraft service was launched in August 1962 with the SR.N2. The eighteen-seat Warden-class SR.N5 and thirty-eight-seat Winchester-class SR.N6 series of multi-purpose small hovercraft were launched in quantity production by 1963 and the larger military SR.N3 craft entered service on loan to the Inter-Service Hovercraft Trials Unit (IHTU) in April 1964. Until the advent of the American-built Landing Craft Air Cushion (LCAC), the 58.5ft 10.9-ton SR.N6 series held the record for the longest hovercraft-type production run, a total of fifty-five craft in several variants.

More astonishing, given the relatively short period of operational experience gained with this type of marine craft, plans were underway as early as 1964 for a giant cross-Channel car and passenger-carrying craft, the Mountbatten-class SR.N4. Model tests in the company's own test tanks and on the local beaches confirmed the practicability of such a vessel and by November 1965 work on the first craft, for British Rail Seaspeed, had started. At 190 tons, 131ft long and capable of a maximum speed of 83 knots (95.5mph), when they entered service in 1968 they were almost fifty times bigger than the SR.N1 completed just nine years earlier. The SR.N4 Mk I craft could carry 254 passengers and 30 cars. Just five years later the Mk III variant of the SR.N4, the 'Super-Four' as it was called, commenced operations, weighing in at 315 tons, eighty times bigger than the first hovercraft in only fourteen years! It had capacity for 418 passengers and 60 cars.

Central to Saunders-Roe's extraordinary development of practical hovercraft was an inspired young engineer, Raymond Leslie Wheeler, who had served his apprenticeship with the company from 1945. Born in Middlesex in October 1927, Ray was educated on the Isle of Wight, attending the Newport County Secondary Grammar School. During his apprenticeship he gained a BSc degree in aeronautical engineering in 1948 from University College, Southampton, following this up with a masters degree in a three-year postgraduate course at Imperial College. These achievements make for a rare distinction, his period of company training constituting what is known as a premium apprenticeship.

Involved in many of the company's design projects undertaken in the 1950s, Ray Wheeler was attached to the hovercraft development team, working on all craft beginning with the SR.N1, in which he had a vital role. By the time that the SR.N4 hovercraft was under development, for which he was also project engineer, he had been promoted to chief structural engineer. Not just confined to his leadership work on hovercraft, Ray was also made responsible in parallel for the Black Arrow rocket programme and was instrumental in ensuring its outstanding success. From 1966 through to 1985 he was made British Hovercraft Corporation's chief designer, having also been promoted to technical director from 1972. From 1985 he became business development director, eventually appointed systems support director for what was by then Westland Aerospace from 1989 through to his retirement in 1991.

In 1995, Ray Wheeler was the beneficiary of a rare distinction when he received the award of Royal Designer for Industry. Where the subject of this book is concerned, it is an almost unique recognition which he shares with Renato 'Sonny' Levi of *Surfury* fame. Since retirement, he has devoted himself to chronicling the company's long and enterprising history and recording the technical and service details of the numerous craft that have been built by the company from the days of Sam Saunders through to the present day GKN Aerospace. Described by himself as a '20th Century Engineer', it could be said that as a man of exceptional engineering vision, he was involved in exploring possibilities that have extended well beyond the millennium.

In effect, viewed in retrospect, the giant SR.N4 Mk III turned out to be the high point for the hovercraft and the British Hovercraft Corporation, as Saunders-Roe (a Westland Aircraft subsidiary since 1959) had become in April 1966. Struggling with the unpredictable and erratic nature of orders and the lack of commitment from the government in the form of demand for military hardware, despite regular high-profile declarations of intent in pre-election statements, the company was obliged to divert into the more consistent manufacture of products for sustained markets, flight-critical components for worldwide aircraft manufacturers. In this, the company, first renamed Westland Aerospace and now known as GKN Aerospace, has been conspicuously successful but in parallel the hovercraft industry, for which so much had been promised, began to wither on the vine.

The potential value of hovercraft business to the UK economy had been clearly demonstrated in both sales of craft to foreign customers as well as in licence technology deals but the company could not sustain itself on an irregular 'feast and famine' order cycle in the absence of support at national level.

There was a new generation of medium-sized, diesel-powered hovercraft (all previous models had been gas-turbine powered) in the pipeline in the form of the AP1-88 and a replacement craft for the cross-Channel service was also in the early planning stage.

As it has turned out, the successful AP1-88 has been built in several versions, initially at East Cowes, although the aluminium hulls of the early craft were fabricated on the mainland, and then taken to the Hoverwork construction plant at the Duver, St Helens, for completion. Developed craft, the BHT-130 and BHT-150, have followed but by the time of their entry into service the involvement of the company which had begun hovercraft development had long ended. After Hoverwork was absorbed by Griffon Hovercraft, the Duver works was closed down and new hovercraft production now takes place in Woolston, Southampton on the River Itchen.

The replacement cross-Channel craft, the BH-88, never advanced beyond the outline concept while a minesweeping variant of the SR.N4 for naval service, the SR.N4 Mk IV, also failed to materialise.

Saunders-Roe and, later, the British Hovercraft Corporation, may have built the world's first hovercraft as well as the world's largest hovercraft but it was not the only Island company to be involved in hovercraft development. Three other local concerns were active during the embryonic stages of the hovercraft industry, designing and building small air cushion vehicles.

The Cushioncraft business at Bembridge, created by Desmon Norman and John Britten of Britten-Norman Aircraft, discussed in the next chapter, produced a sequence of craft types, the CC-1 through to the CC-5 and CC-7 (there were no CC-3 and CC-6 machines, the latter a planned joint venture with Vosper Thornycroft, as they did not proceed beyond the concept stage), each an improvement on the previous craft as handling experience was gained with successive machines. The CC-1 was second only to the SR.N1 in being the first hovercraft in the world, its maiden test hover having occurred in June 1960. The Cushioncraft concern was acquired by British Hovercraft Corporation in January 1972 which then built two CC-7 hovercraft for the British Army.

The boatyard at J. Samuel White in Cowes also built a test hovercraft, the HD-1, for Hovercraft Development Limited and at Fishbourne on Wootton Creek, Enfield Marine, which had produced race-winning powerboats to Don Shead's designs, embarked upon its own bid to enter the hovercraft market with its EM.1 and EM.2 designs. In the event, only a single EM.2 craft was produced.

The smaller EM.2 was only intended to be a test bed for the planned, larger EM.1 but it proved to be a worthwhile logistics craft in its own right, boosted by a complimentary evaluation commissioned by the United States Navy, in which it was compared with other similar-sized air cushion craft. In the event, though, the EM.1 never materialised and, despite the favourable plaudits it had received, orders for the EM.2 were not forthcoming.

Arising from company's innovative record, Saunders-Roe was awarded the contract from NRDC to construct the world's first practical hovercraft, the SR.NI, here seen during assembly in the Columbine hangar in early 1959. Completed in only eight months, it was powered by an Alvis Leonides 450hp engine. (Saunders-Roe)

The completed SR.N1 in hover mode. On 25 July 1959, the fiftieth anniversary of Louis Blériot's first flight across the English Channel, the SR.N1 made the first crossing of the Channel from Calais to Dover by hovercraft. The SR.N1 in its final form is now held in the Large Object Store of the Science Museum at Wroughton, near Swindon. (Saunders-Roe)

The 65ft-long SR.N2 hovercraft nearing completion in the Columbine hangar in 1961. The world's first purpose-built passenger-carrying hovercraft, the inaugurated SR.N2 a service between Ryde, Isle of Wight, and Southsea, Hampshire in August 1962. (Saunders-Roe)

Completed in November 1963, the 37-ton, 77ft SR.N3 military hovercraft was the world's largest hovercraft for the next four years. It was conceived as a trials vehicle for the Inter-Services Hovercraft Trials Unit (IHTU) to evaluate potential roles for this type of waterborne craft with the armed forces. Subjected to underwater mine explosion tests in 1974, the SR.N3 demonstrated that, with their air cushion, hovercraft were largely impervious to such weapons, suiting them to a mine countermeasures role. (Saunders-Roe)

145

Hovercraft development was rapid and within four years a series of practical craft were in production, the SR.N5 Warden-class and SR.N-6 Winchester-class types powered by Rolls-Royce Marine Gnome gas turbines. Designed for sheltered, short duration passenger ferry services and coastal logistics work over mudflats, marshes and across river estuaries, they were also produced in military variants. In retrospect, they were the company's most successful designs with more of these craft built than any other Saunders-Roe/BHC hovercraft, a total of sixty-nine. The photograph shows an SR.N5 operating over ice. (Saab A/B)

BHC's ambition was to build much larger craft for cross-Channel service, able to operate in rougher sea states. The result was the giant SR.N4 Mounbatten-class passenger and vehicle craft. A model of the planned SR.N4 is put through its paces from an Island beach probably in 1965. The model shows an early configuration of the design when cross-Solent services were also a target market. (Saunders-Roe)

DAILY EXPRESS

Tuesday June 4 1968

BY HOVERFAST
—across the Channel!

Ready to conquer the Channel, British Rail's majestic S.R.N.4 hovercraft skims over choppy seas during trials

Above: The front-page headline from the *Daily Express* of 4 June 1968, proclaiming the inauguration of the first cross-Channel hovercraft service from Dover to Calais, opened with the SR.N4 Mk I *Princess Margaret*, GH-2006. A second service from Ramsgate to Calais followed with SR.N4 Mk II craft. (Author's collection)

Opposite bottom: The only hovercraft completed by Enfield Marine was the 40ft EM.2 logistics craft completed in January 1972, initially intended as a proving vehicle for the larger 72ft EM.1 freight transport that the company also intended to build. Despite a favourable evaluation by the United States Navy in competition with other similar-sized hovercraft, including the Bell SK-5 based on the Saunders-Roe SR.N5, no real interest followed. The simple and rugged EM.2 was one of the first hovercraft to have ducted propellers and to be powered by petrol engines, twin AEC V8s each producing 420hp. (United States Navy)

Above: The two original SR.N4 Mk I craft were enlarged into the Mk III or Super-4 variant which remains to this day the world's largest commercial hovercraft at 320 tons and measuring 187ft in length. On 14 September 1995 the Super-4 *Princess Anne*, GH-2007, set the record for the fastest ever commercial crossing of the Dover Straits, taking just 22 minutes to cover the 26 miles at a speed of 61.7 knots (70.9mph). This photograph gives a good indication of the immense size of these craft. (British Hovercraft Corporation)

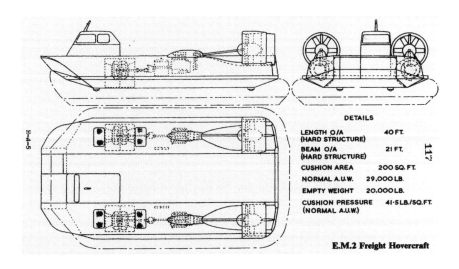

DETAILS

LENGTH O/A (HARD STRUCTURE)	40 FT.
BEAM O/A (HARD STRUCTURE)	21 FT.
CUSHION AREA	200 SQ. FT.
NORMAL A.U.W.	29,000 LB.
EMPTY WEIGHT	20,000 LB.
CUSHION PRESSURE (NORMAL A.U.W.)	41·5 LB./SQ. FT.

E.M.2 Freight Hovercraft

John Britten and Desmond Norman also recognised the potential of amphibious hovercraft and created a subsidiary of Britten-Norman called Cushioncraft. The CC-1 was its first design, followed by other experimental craft up to the CC-5, seen here, completed in March 1966. Involved in this enterprise was the talented designer John Ackroyd, of whom more later. (Cushioncraft)

The BH.7 was a medium-sized hovercraft intended for military roles drawing on the experience with the SR.N3. It was the first craft to have a British Hovercraft Corporation designation. Conceived for either logistical transport of troops and vehicles or coastal patrol, this craft was later fitted with a Plessey Marine sonar and a Racal positioning systems navigation system for mine countermeasures trials with the Naval Hovercraft Trials Unit (NHTU). The photograph shows the BH.7 Mk II as modified with a clamshell bow door that was fitted in 1973 approaching the beachhead at HMS *Daedalus*, Lee-on-Solent. (Graham Stevens)

Evolved from Cushioncraft's CC-5 was the CC-7, a slightly larger craft that had production potential. After the company was absorbed into British Hovercraft Corporation in January 1972, two of these craft were completed for the British Army. Unlike other amphibious hovercraft, the Cushioncraft CC-7 had flexible, inflatable side decks, which reduced weight. Nevertheless, with a single United Aircraft of Canada ST6B gas turbine providing both lift and propulsion, the maximum speed was only 35 knots (40.2mph). (British Hovercraft Corporation)

An integral dimension of BHC's business activities was the licensing of its technologies to foreign customers, a valuable export business. One such arrangement, concluded through the USSR's licensing agency Licensintorg Resource Sciences, granted a licence for the design, testing and development of hovercraft centrifugal lift fans. Besides the transfer of company expertise in this area, the deal also involved the supply of SR.N4 lift fans that were incorporated in the Soviet Navy's very large Aist-class hovercraft, an example of which is shown here. (Author's collection)

In this case the licensing of BHC technology concerned inflatable rubber-shirt systems based on those developed for the SR.N4 Mk III craft. Under a contract with the United States Navy, finished skirts as well as the detailed designs for these structures were supplied for the Landing Craft Air Cushion (LCAC) programme. Now the most produced hovercraft of all time, a total of ninety-one LCAC craft have been constructed for service with all classes of US Navy logistic ships. (United States Navy)

The first new generation of medium-sized passenger hovercraft, the diesel-powered AP.1-88-80 with twin ducted propellers, entered service with Hovertravel in March 1983. Hovertravel's associate company, Hoverwork, collaborated with Westland Aerospace formerly BHC in the development of the type. Named *Tenacity*, its aluminium hull had been constructed by Allday at Gosport and fitted out in Hoverwork's workshop in Bembridge Harbour. Measuring 70ft in length, the AP.1-88-80 was powered by Deutz Diesels rated at 328bhp, a much quieter option than the Rolls-Royce Gnome gas turbines which had been used on SR.N6 craft. (David L. Williams)

Two AP.I-88-80 craft, painted navy grey, were chartered to the United States Navy for crew training for the larger LCAC craft then coming on stream. (Westland Aerospace)

flotaload
hoverpallets

'FLOTALOAD' hoverpallets are simple air-cushion load platforms designed to facilitate the storage and movement of
goods in factories, warehouses, aircraft and ships. Being
independent of costly fixed installations, they enjoy
unrivalled freedom of movement and their high manoeuvrability
is of particular value in confined spaces.

They are light yet robust in construction and can easily be
moved from location to location.

They are economic and simple to operate and the only accessory
equipment needed is either an industrial airline or a mobile
compressor. Maintenance is greatly simplified by the absence
of mechanical parts.

Above all, they have a high work capacity. Current types can
carry up to 5 tons and manpower requirements are reduced
to a minimum.

'FLOTALOAD' hoverpallets can either operate as independent
units or be integrated into existing handling systems and they
have been designed to work in conjunction with containers
and all standard sizes of fork lift pallets.

Apart from savings in manpower and time, 'FLOTALOAD'
hoverpallets offer further advantages which enhance their
operational flexibility. By spreading loads over the area of
the cushion, they are able to operate safely over weak floors
which could not otherwise withstand the concentrated axle
loading of conventional wheeled vehicles and the absence
of wheels reduces wear on floor surfaces. In addition, their
smooth progress makes them ideal for the movement of delicate
instruments and machinery.

Two standard sizes of 'FLOTALOAD' hoverpallets, capable of
carrying loads of one ton and five tons, are currently available
and new sizes are being added to the range.

We can also advise on specialised applications.

FLOTALOAD is designed and manufactured under British Patent
No. 885235 and corresponding foreign patents Other patents pending.

'Whilst it is believed that the contents of this document are correct at the
time of going to press it must be appreciated that this brochure is for
information purposes only and that it does not form the basis of any Contract
with The British Hovercraft Corporation Limited.

British Hovercraft Corporation is a subsidiary of Westland Aircraft Limited.

british hovercraft corporation limited

Publication No. SP 1584 Issue 1 May 1967 Printed in England

BHC pursued numerous innovative applications of the hovercraft technology it had developed
for other vehicular, logistics and manufacturing purposes, exploiting both the principles of air
cushion lift and its flexible rubber-skirt systems, designed to provide peripheral containment
of pressurised air. Its trademark 'Flotaload' hoverpads and hover pallets were devised to
facilitate the movement of heavy loads around factories, warehouses and airports, and even
in the holds of ships. (John Fisher collection)

air cushion
heavy load
transporter

This vehicle has been developed to facilitate the movement by road, of heavy industrial equipment such as electricity transformers, by distributing more evenly the load imposed on bridges.

This is achieved by using air-cushion equipment located under the central load-carrying platform of the transporter.

The development of this equipment was prompted initially by the Central Electricity Generating Board, which is constantly faced with route-planning problems caused by the high weights of laden transporters.

Transformer units now going into service weigh between 195 and 250 tons and on occasion the C.E.G.B. has been involved in the heavy expense of strengthening and even rebuilding bridges to accept these loads when no alternative route has been available.

The use of air-cushion equipment, however, provides a practical and economic alternative.

On its first commercial trip in February, 1967, the converted transporter had to cross the Felin Puleston Bridge near Wrexham in North Wales, carrying a 155-ton transformer. Using the air-cushion equipment, it was possible to reduce stress on the bridge by 70 tons, bringing it well within the load limitations imposed by the Ministry of Transport.

Without the air-cushion equipment, it would have been necessary in this instance to rebuild the bridge at a cost of £30,000. This saving represents half of the total development cost of the transporter.

The cushion of air itself is contained in a curtain of nylon/neoprene, stretched as a seal beneath the load platform and formed into convolutions about its periphery. It is protected from wear through contact with road surfaces, by steel plates. Any unevenness of road surfaces causes the convolutions to flex but allows the plates to remain in contact with the road, thus maintaining the air seal.

The cushion is capable of lifting 155 tons, at which point the pressure reaches 5.4 p.s.i.g. The power to generate and maintain this pressure is provided by four Rolls Royce B.81 petrol engines, delivering 235 h.p. at 4,000 r.p.m., which are housed in a sound-proofed vehicle at the rear of the transporter. Air is fed to the cushion by ducting.

To avoid possible damage to bridges due to sudden fluctuations in cushion pressure, this is controlled automatically to very close tolerances. Should one power source fail, it is automatically sealed off. In normal operation, 25% of the air supply is blown to waste but in the event of a power unit failure, this is automatically diverted into the cushion, thus maintaining a constant cushion pressure. The whole system is sound-proofed to satisfy stringent limitations on noise level laid down by the Ministry of Transport, and the air cushion does not affect other road users. When the cushion is not needed, the curtain can be quickly detached and folded for stowage.

Whilst it is believed that the contents of this document are correct at the time of going to press it must be appreciated that this brochure is for information purposes only and that it does not form the basis of any Contract with British Hovercraft Corporation Limited.

British Hovercraft Corporation is a subsidiary of Westland Aircraft Limited.

Publication No. SP 1517 Issue 1 May 1967 Printed in England

 british hovercraft corporation limited
YEOVIL ENGLAND

Another important implementation of air cushion technology by BHC took the form of the patented Heavy Load Transporter, developed with the Central Electricity Generating Board (CEGB) in mind for the conveyance by road of large, very heavy objects such as power station transformers. It comprised a towed hover platform which supported the load, connected to articulated units with generators and pumps to supply air for the lift cushion. This device permitted extreme loads to be taken over weight-restricted roads and bridges without causing structural damage. (John Fisher collection)

Above: An enduring hovercraft customer has been the Canadian Coast Guard for whom the AP.1-88-200 and -400 well-deck versions were evolved for ice breaking and navigational aid (navaid) maintenance roles in the St Lawrence Seaway and off the coast of British Columbia. Here, the first of these craft, the *Waban Aki*, is seen under construction in the Hoverwork assembly building at the Duver, St Helens. (Nick Henry)

Opposite bottom: The early years of hovercraft development were characterised by extraordinary optimism and conjecture as to the future possibilities of this revolutionary form of transport. Some extremely ambitious plans were hatched, notably for very large, transoceanic craft such as ocean passenger ferries and hover-freighters up to 3,000 tons. Equally, schemes for large vehicles for naval operation soon featured on the drawing board, including 90-knot frigates for the Royal Navy to be built in the Falcon Yard. Possibly the most extraordinary vessel then under consideration was this 1,500-ton, 288ft nuclear-powered hovercraft mini-aircraft carrier, dating from 1960. (Saunders-Roe, courtesy of Ray Wheeler)

Above: The basic AP.1-88 craft design has itself also gone through a substantial enhancement programme, emerging in a totally redesigned form as the much-improved Fisheries & Oceans, Canada BHT series of hovercraft. There are passenger variants, a new search and rescue well-deck craft as well as logistics vehicles. Designed by Hoverwork and built at St Helens, the BHT-150 *Mamilossa*, is shown in her Canadian Coast Guard livery. The craft measures 93.5ft by 39.3ft and has four Caterpillar twelve-cylinder diesels, each producing 1,125bhp for a maximum speed of 50 knots (58.0mph). (Fisheries & Oceans, Canada)

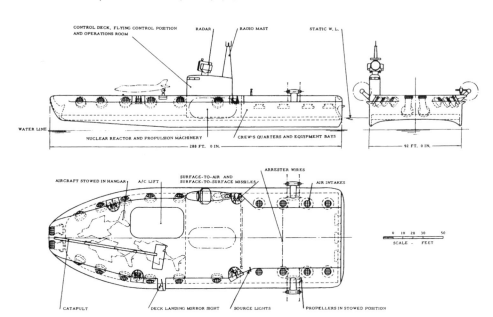

1,500-TON NUCLEAR-POWERED HOVERCRAFT AIRCRAFT CARRIER

Above: Once viable hovercraft operations were up and running, BHC had to give thought to the replacement craft that would eventually be needed. The AP.1-88 had fulfilled this requirement admirably where the SR.N6 craft were concerned, introducing important improvements like greater capacity, enhanced fuel efficiency and quieter operation. For the Channel run, the BH.88 was devised as the future replacement for the Super-Four craft. However, in the very year that work commenced on this design, work also started on the Channel Tunnel, which opened six years later on 6 May 1994. Offering an equally fast means of transit, unaffected by weather conditions, it immediately undermined the hovercraft's market. The two Super-Fours, *Princess Anne* and *Princess Margaret*, were retired in 2000 without ever having been replaced by superior craft. This painted impression of the BH.88 is by Peter Thorne, an artist who worked in the company's Technical Publications Department. (Westland Aerospace)

8

FAMILY OF ISLANDERS

The infant Spartan Aircraft Company, originally established in 1928 by Oliver E. Simmonds, was acquired in 1930 for Saunders-Roe by the Aircraft Investment Corporation through a refinancing arrangement. The purpose of the acquisition was in part to permit the construction of airframes to Spartan's designs in the Cowes factory as a means of supplementing flying boat assembly, work which at the time was at a low ebb. However, the first attempt on the Island to create and operate a small but practical light commuter or transport aircraft also emerged through this subsidiary concern. This was the Spartan Cruiser, introduced in 1932.

Spartan's initial designs had been for club flying, joy riding, competition and air taxi machines in the shape of the Spartan Arrow and Three-Seater Mks I and II, of which a total of ninety were completed – thirty-seven at the Somerton works in Cowes following the takeover.

Meanwhile, around the time of the takeover, Saunders-Roe had begun developing a high-speed mail-carrying aircraft for Empire routes, designed to have a range of 1,000 miles with a payload of 1,000lb. This was the Saunders-Roe three-engined A.24, designed by E. W. Percival, only one of which was built, to be subsequently transferred under the Spartan banner as the Spartan Mailplane. However, the Mailplane failed to attract Air Ministry interest and as no other market materialised for the machine it was adapted as the design prototype for the subsequent A.24M Cruiser, a small civil transport plane type-approved in three marks. The Cruiser was specifically intended for domestic and near continental air routes, including to the Channel Islands. Its passenger-carrying capacity was up to a maximum of eight persons with a two-man crew.

An operating company, Spartan Airlines, was also set up by Saunders-Roe but the Cruiser also saw service with United Airways Limited, British

Airways Limited and Northern & Scottish Airways, all functioning as part of the Railway Air Services network. A total of seventeen Cruisers were completed by the time production ended and Spartan Aircraft was wound up in 1935. Sales had been actively promoted at home and abroad but with the new London flying boats on the order book, factory effort had to be geared up for them as well as concentrated on completing the residual work on the A.29 Cloud amphibians for the RAF. The various operators that had deployed Cruisers were progressively amalgamated into a single concern which eventually formed an integral part of British Overseas Airways Corporation (BOAC), the predecessor of today's British Airways.

The fact was that back in the 1930s air travel was an expensive and exclusive form of transport and inevitably, despite a quite extensive domestic route network, there was not a sufficient level of patronage to drive production orders for substantial numbers of new aircraft. Nonetheless, the Spartan Cruiser constituted an important stage in the evolution of commuter aircraft and mass air travel, as one of few such aircraft in existence at the time of its introduction.

Thirty or so years later, two enterprising design engineers revived the concept of a small commuter and general purpose utility transport with a measure of success that must have been beyond anything they could remotely have imagined. They were Forester Richard John Britten (known as John) and Nigel Desmond Norman (known as Desmond).

John Britten was born in Windsor in May 1928 and Desmond Norman was born in London in August 1929. They met as young men during their engineering apprenticeships at the de Havilland Technical School and immediately struck up a friendship having a shared passion for aviation and sailing. Following national service, their partnership was cemented with the construction of the first aircraft to their own designs, the BN-1F Finibee now preserved in Southampton's Solent Sky Museum.

With two more partners, Jim McMahon and Frank Mann, Crop Culture (Aerial) Ltd was formed operating crop-spraying aircraft and, following the invention of the rotary atomiser by Edward Bals, this led on to the formation of Micronair Limited, an Island-based subsidiary company specialising in spraying equipment. The crop-spraying operations generated the finances required to permit the partners to resume the design and construction of more aircraft, their main ambition. In 1955, Britten-Norman Limited was set up expressly for this purpose. However, attention was diverted to hovercraft development in response to an idea germinated by Elders & Fyffes that such craft could provide the ideal form of transportation for conveying bananas from the plantations in the southern Cameroons to the docks for loading aboard ship.

Work began at Bembridge in 1960 on the first hovercraft, the CC-1, by the Britten-Norman hovercraft division which a year later was reconstituted as the subsidiary concern, Cushioncraft Limited. While hovercraft development continued through several design variants, this was to some extent a deflection from the main goal which was to create a rugged and durable inter-island passenger and light utility aeroplane, to some extent, in retrospect, reviving the concept pioneered with the Spartan Cruiser. Work on the new machine began in earnest in 1963 under the designation BN-2, the small design team supplemented by designers and draughtsmen seconded from F.G. Miles of Shoreham, among them Denis Berryman who subsequently joined Britten-Norman as chief designer. On 13 June 1965, the prototype aircraft, G-ATCT, made its maiden flight over the Island and on 15 August 1966, just a few months before the inaugural flight of the production prototype, G-ATWU, the name 'Islander' was appropriately and popularly selected for the BN-2. From this point, progress gathered pace rapidly.

A new production factory was opened at Bembridge that December to provide capacity to handle the growing order book. The first production aircraft was delivered on 13 August 1967 and by September 1969 deliveries had reached 100. It was busy times for the small company. A second new aircraft, the BN-3 Nymph, a four-seat, high-wing monoplane made its first flight on 17 May 1969, the BN-2 Defender was added to the series in May 1970 and the three-engined BN-2A Mk III aircraft, for which the name 'Trislander' was adopted in January 1971, flew for the first time in July 1970. However, 1971 proved to be a year of mixed fortunes. The first Trislander was delivered on 29 June but barely four months later, having suffered its first bout of financial difficulty, the business was placed in receivership on 22 October.

Fortunately, it was able to continue as a going concern under the name Britten-Norman (Bembridge) Limited, to be taken over in August 1972 by Fairey S.A. in the form of the holding company Fairey Britten-Norman. This arrangement initially worked well, even though production of some of the aircraft was transferred to Gosselies in Belgium, adding to the already out-sourced production in Romania and at the British Hovercraft Corporation in East Cowes where ultimately around 500 and 363 aircraft respectively were constructed. At this time, feasibility studies were commenced for a fourth aircraft type, the larger, four-engined BN-4 Mainlander.

Deliveries of the BN-2 Islander reached 500 on 16 August 1974 and, a month later, the British production record for a non-military aircraft was achieved after 549 deliveries. Despite this, John Britten and Desmond Norman, who had held board positions in Fairey Britten-Norman, left the company in February 1976 although they continued to serve as consultants. A year and a half later, new financial difficulties beset the company as a result of stockpiling of undelivered airframes in Belgium and on 3 August 1977 the receivers were called in again. Rescued a second time, an offer from the Swiss company Pilatus was accepted and by January 1979 the company was re-registered under the name Pilatus Britten-Norman.

Meanwhile, new variants of the BN-2 had come off the drawing board and gone into production: the BN-2B Islander II and the BN-2T turbine Islander which had its maiden flight on 2 August 1980. The production milestones continued to clock up, and just under two years later, on 7 May 1982, the 1,000th BN-2 delivery was reached. CASTOR, AEW and MSSA derivatives of the Defender type took to the sky between 1984 and 1994 and the latest in this family of aircraft, the BN-2T-4S Defender 4000 entered service in 1996.

In just over a quarter of a century, 1,280 of these ubiquitous, multi-role aircraft had been produced, a phenomenal record for an aircraft of this type and a huge tribute to Isle of Wight industry and to the foresight and energy of the Islander's originators in the pursuit of their dream.

John Britten was unable to pursue new objectives for very long after leaving the company he had helped to create, passing away in July 1977. Aided by Denis Berryman, former chief designer at Britten-Norman, and Maurice Brennan, former chief designer at Saunders-Roe, John Britten *did* establish Sheriff Aerospace to produce the Britten Sheriff aircraft, a twin-engined low cantilever-wing monoplane designed for touring or training purposes. The construction of the single aircraft of this type, G-FRJB, reached a fairly advanced state only to be abandoned incomplete, the untimely death of its originator contributing to the termination of the project.

His former partner, Desmond Norman, went on to introduce two exciting new aircraft by his new company NDN Aircraft Limited, established at Sandown in 1976. It later became the Norman Aeroplane Company when its design office and manufacturing base were moved to Cardiff Airport in July 1985, induced into going there by favourable industrial enterprise start-up grants offered by the Welsh Development Agency.

The first of Desmond Norman's new aircraft was the NDN-1 Firecracker, the design for which his new company had been primarily created. It was a tandem-seated, single-engined monoplane military trainer, powered initially by a Lycoming O-540 piston engine at the time of its first flight on 26 May 1977. Later, it was adapted as a turbo-prop variant, the NDN-1T with a Pratt & Whitney PT6 engine. First flown on 1 September 1983, the turbine-engined model was capable of a maximum speed of 198 knots (228mph). However, only four machines were built, its prospects dealt a major blow when it was not selected as the replacement for the RAF's Jet Provost trainers.

It can only be assumed that there were designs NDN-2 through to NDN-5 but they did not materialise and it is not known what type of aircraft they would have been. The next and last aircraft to emerge from the NDN stable was the NDN-6 Fieldmaster, a low-wing monoplane with a fixed tricycle undercarriage and single 750shp Pratt & Whitney PT6A turbo-prop engine, designed specifically for agricultural use, spraying crops. The maiden flight of the prototype took place at Sandown on 17 December 1981. The Fieldmaster was also adapted as a specialised fire-fighting aircraft, with the name Firemaster, but success with this new venture again eluded Desmond Norman and only ten aircraft in total were completed. It turned out to be a mute swansong to a career which had enjoyed immense success with the BN-2 Islander.

Spartan Aircraft's metal fuselage, wooden wing Cruiser was one of several aircraft designed to open commuter services as domestic air routes were established throughout the UK in the 1920s and 1930s. Regular flights were offered from Cowes and Ryde to Shoreham, Croydon, Castle Bromwich and elsewhere. This is a six-seater Cruiser Mk II, G-ACDW, in the colours of Spartan Air Lines. (Milestone)

This advert for Spartan Aircraft from 1933 gives Cowes as the company's address. In fact, while the aircraft were assembled at Somerton, the design office, integrated into that of Saunders-Roe was in East Cowes, in the still existing Esplanade building. (Aviation Ancestry)

SPARTAN CRUISER

7-8 seat cabin monoplane for air-line services. Guaranteed to climb to 5,000 ft. and maintain that height on any two engines.

SPARTAN AIRCRAFT LTD., COWES, ISLE OF WIGHT

This is an improved Mk III Cruiser, G-ADEL, flying over East Cowes. The Mk III variant had capacity for eight passengers, comparable to the later Britten-Norman BN-2 Islander introduced thirty years later. Powered by three de Havilland Gipsy Major engines, the later Cruisers had a range of from 550 to 600 miles at a cruising speed of 118mph. A total of seventeen Spartan Cruisers were built. (Milestone)

Above: The concept of a utility commuter aircraft was revived in the 1960s by Britten-Norman, a company set up in the mid-1950s by John Britten and Desmond Norman, which became the island's most prolific and most successful aircraft manufacturer. The sole BN-1F Finibee, the first aircraft designed and built by the partnership of John Britten and Desmond Norman was assembled in the original aircraft workshop at Bembridge in 1951 and is now preserved in the Solent Sky Museum, Southampton. (Arthur Ord-Hume)

Opposite bottom: Britten-Norman's outstanding success was with the twin-engine eight-seat BN-2 Islander series, initially powered by piston engines, which enjoyed an incredible production run from its introduction in June 1965. Fuselage length was 36ft and wingspan was 49ft. Here the two men who were the inspiration behind the Islander, John Britten and Desmond Norman, are seated, in the full-scale wooden mock-up in late 1964, along with Ernie Perkins and Denis Berryman. (Peter Gatrell)

Above: Designed by a team led by chief designer Denis Berryman, formerly of F.G. Miles, this is the prototype machine, G-ATCT, seen during assembly at Bembridge in early 1965. The BN-2 Islander production run, in variants with piston-engines or turbo-props, resulted in more than 1,120 aircraft, a record for a British civil aeroplane. (Bob Ward)

The piston-engined prototype Islander G-ATCT (BN-2-001) in flight over the Needles on 17 December 1965. By then its original Rolls-Royce Continental engines had been replaced with AVCO Lycoming 0-540 engines and the wingspan had been extended by 4ft. (Peter Gatrell)

The first turbo-prop Islander BN-2T with Rolls-Royce Allison 250-B17C engines made its maiden flight on 2 August 1980. Full CAA certification was achieved a year later. To describe the Islander as a utility aircraft was an understatement given its application to passenger service, coastal patrol, air ambulance, executive transport, parachute training, logistical supply and many other roles. (Britten-Norman)

For paramilitary and defensive patrol work, the BN-2B Defender variant was introduced in June 1971. Supplied initially with piston engines, later Defenders were powered by turbo-props. (Britten-Norman)

In 1977, Miles Dufour fitted Islander G-FANS with Dowty Rotol ducted fans. Intended as a demonstrator for this alternative to conventional propellers, the aircraft exhibited excellent performance with a significant reduction in noise. (BNAPS Archive)

A Britten-Norman advert from May 1971 promoting the BN-2 Defender type. This variant was also advertised as the Maritime Defender, focusing on the need to patrol offshore assets such as the oil and gas platforms in the North Sea. (Aviation Ancestry)

A view of the inside of the new Britten-Norman works at Bembridge, opened in 1966 and described as 'Europe's finest light aircraft factory'. Production was notably intense at this time when Islander orders were at a peak. To cope with the demand, airframe assembly was sub-contracted to the British Hovercraft Corporation in East Cowes and, under a licence deal, to Romaero SA of Romania. The Fairey Group which later acquired Britten-Norman also had some production carried out at its Gosselies plant in Belgium. (Britten-Norman)

Above: A significant development of the Islander in 1971, to provide greater power, range and capacity, was the three-engined BN-2A-III Trislander which had a third engine mounted in the tailplane. A total of seventy-two of these eighteen-seat workhorse commuter aircraft were built. They gave sterling service over working careers of thirty years and more. This is a BN-2A-III-2 long-nose version in the colours of Inter Island Airways. (Mike J. Hooks)

Opposite bottom: A smaller aircraft developed by John Britten and Desmond Norman was the BN-3 Nymph, a light two-seat monoplane, 25ft long with a 40ft wingspan and hinged wings which folded back against the fuselage. The first Nymph, G-AXFB, flew at Bembridge for the first time on 17 May 1969. Only two of these aircraft were built but the design was later revived and revamped as the Norman Aeroplane Company's Textron Lycoming-powered NAC Freelance. (Ivan Berryman)

Above: Another Islander adaptation was this amphibious conversion to aircraft G-AVCN, undertaken in 1974 for the Philippines Aerospace Development Corporation. However, the floats imposed an unduly heavy weight penalty and the concept had to be shelved pending the development of a larger, more powerful variant of the BN-2 family. (BNAPS Archive)

Yet bigger aircraft were planned by Britten-Norman during the Fairey Britten-Norman era. Among them was the BN-4 Mainlander, shown here. Measuring 46ft in length with a wingspan of 60ft, the four-engined aircraft had capacity for twenty-one passengers. Another project under consideration was a three-engined short take-off and landing (STOL) prop-jet aircraft in three variants: a 100-passenger short-haul airliner, a vehicle carrier and a logistics transport for bulky cargoes. Financial difficulties, takeovers and the departure of the two main players from Britten-Norman prevented progress with these concepts. (Ivan Berryman)

The Islander was identified as an aircraft particularly well suited to both long-range battlefield radar surveillance and early warning radar surveillance roles. Therefore, evaluation trials were arranged with equipment installed aboard two former piston-engined aircraft modified to turbo-prop configuration. Shown here is G-DLRA fitted with Ferranti Defence Systems' Corps Airborne Stand-off Radar system (CASTOR). Test flights commenced on 17 May 1984. (Britten-Norman)

Simultaneously, aircraft G-TEMI was adapted as an AEW Defender with a Thorn-EMI 360-degree radar antenna installed in a projecting nose cone. This was capable of detecting fast-moving objects at both high and low levels. Trials with this variant began on 18 July 1984 lasting almost three years. (Britten-Norman)

When renewed financial difficulties led to the takeover of Britten Norman by Pilatus of Switzerland, John Britten and Desmond Norman eventually went their separate ways. Desmond Norman created NDN Aircraft, based at Sandown, which produced a number of interesting aircraft, among them the NDN-1 Firecracker for pilot training offered with two engine options. (Mike J. Hooks)

The NDN-6 Fieldmaster could be equipped for two different roles, either with water scoops for fire-fighting, as shown here, or with Micronair Rotary Atomiser equipment – a product developed by another Island-based Britten-Norman subsidiary – for crop spraying. (Mike J. Hooks)

Arising from the trials with the CASTOR Islander and AEW Defender, a further surveillance variant was developed, the Multi-Sensor Surveillance Aircraft (MSSA) – see the colour section. Experience with the MSSA aircraft revealed a need for more working space within the fuselage. As a consequence, another Islander variant, the BN-2T-4S Defender 4000, was developed. The fuselage length was increased by 4ft to 40ft; it had a larger wing and longer nose and a cockpit with better all-round visibility. The first Defender 4000 made its debut on 17 August 1994. (BNAPS Archive)

9

MAN, MACHINE & THE ELEMENTS
A Unique Contribution

This chapter breaks the mould of the previous chapters by focusing on one individual, John Ackroyd, whose contribution to the Isle of Wight's industrial and manufacturing heritage does not readily fit into any one single category because it spans many diverse products and vehicles. However, despite having been involved in so many incredible, groundbreaking and high-profile projects, his achievements have not, perhaps, received recognition nationally to the extent they deserve and, despite his name being known by many locally, beyond the boundaries of the Isle of Wight he is in a sense something of an enigma other than to those who have worked with him.

He continues to exude passionate enthusiasm and a strong sense of commitment to the projects he is engaged in, as he always did. Yet he is a rather modest man and perhaps as a consequence his fundamental role in several of the greatest record-breaking accomplishments of recent times has gone, to some extent, un-acclaimed.

John Ackroyd (he likes to be known as 'Ackers') was born in India in 1937. He returned to the UK as a lad of 7 and completed his schooling and higher education here before taking up an apprenticeship at Saunders-Roe. It was a learning experience which he holds in the highest regard and he willingly concedes that much of what followed owes a great deal to the quality of the training he received at the Osborne Apprenticeship School under the guidance of Victor Stephenson (popularly known as V.T.) and his staff.

The interesting thing about what followed is that, ever enthusiastic and confident about what he was capable of doing, he invariably volunteered his services to undertake what were truly challenging commissions, often in areas where he had little or no previous experience. As he says, he drew on the broad knowledge base gained during his apprenticeship. That, together with his evident enthusiasm for a challenge, his disciplined approach to engineering and his practical skills, has comprised his design tool set. The results speak for themselves.

Described as a design engineer extraordinaire, John Ackroyd cut his teeth working at Saunders-Roe up to 1960 on the SR.53 and SR.177 mixed-power interceptor fighters. Contract work with a range of companies on diverse vehicles followed until, in 1965, he joined Cushioncraft at Bembridge supporting design work on early hovercraft, which continued for the next three years. More contract employment followed in Germany until John was invited by Giannis Goulandris in 1971 to join Enfield Automotive at Somerton and design what became the first British production electric car, the Enfield 8000 'city' car, one of the first such vehicles in the world.

Apart from a short break working on Audi car designs when Enfield Automotive closed due to an industrial dispute, John resumed work on the Enfield 8000 and the related Miner and Bikini models in Greece. The basic vehicles were assembled in Syros and then shipped back to the Isle of Wight for fitting out and completion. When production came to an end in 1976, John next responded to an advertisement calling for an engineer to design a 650mph land speed record car, hardly a commonplace vacancy notice or, indeed, an everyday challenge. Not that this in any way deterred John Ackroyd.

The result was his greatest and most amazing creation, the *Thrust 2* record-breaking jet-engined car sponsored by Richard Noble and built in a rundown workshop at Fishbourne, Isle of Wight. It was a project that occupied him over the next eight years, through rough and smooth as funding issues and test setbacks were repeatedly encountered.

The completed car was rolled out in 1981 and that year, after proving runs at RAF Leconfield, it took the UK land speed record at Greenham Common at 248mph. With additional streamlining, *Thrust 2* was taken to the Bonneville Salt Flats in America in the following year where it raised its maximum speed, first to 300mph and then to 400mph but the track surface was not conducive to achieving higher speeds with the wheels then fitted to the car. In September 1982, the first attempt to break the existing world record was made at the Black Rock Desert, Nevada. A speed of 590mph was achieved; just 32mph short of the target, but the onset of bad weather brought a premature end to the effort. Further frustration was experienced when, in May 1983, the team again returned to the United States only to be hampered by more inclement weather. But, that same year, on 4 October, *Thrust 2* successfully completed the mandated two runs within 1 hour over the measured mile to break the record, achieving an average speed of 633.468mph. The record was back in British hands!

With the break-up of the *Thrust 2* team, John Ackroyd embarked on new challenges, first working at Sandown airport for ARV Aviation, a company set up by Richard Noble to produce the ARV Super-2 light aircraft, the letters 'ARV' reflecting its 'Air Recreational Vehicle' concept. As one of designer Bruce Giddings' team, John was responsible for the aircraft's tail end. Then, from 1985, he began designing pressure capsules for a sequence of distance, endurance and altitude record-breaking balloon expeditions, in several of which Sir Richard Branson participated.

Again John Ackroyd's change of direction arose as a response to a newspaper advert, the first of these schemes being for the *Endeavour* super-pressure helium balloon which was intended to make a circumnavigation of the globe in the southern hemisphere, from and to Australia. A succession of other balloon capsule design challenges followed: the hot air balloon *Virgin Atlantic Flyer* in 1987 which extended the distance record from 900 to 3,075 miles; the *Stratoquest* hot air balloon which broke the altitude record at 64,997ft (11 miles) in 1988; the *Virgin Pacific Flyer*, another hot air balloon which, in 1991, again eclipsed the distance record at 6,700 miles; the ill-fated *Earthwinds* helium balloon which attempted a global circumnavigation in 1993; and the *Virgin Global Challenger* balloon, a de Rosier hybrid craft combining hot air and a gas bag, which was also unsuccessful in its attempted round-the-world flight in 1998.

John Ackroyd has also continued with his involvement in land speed record-breaking projects, notably with Rosco McGlashon, supporting the design work on the 55ft 200,000hp *Aussie Invader 5R*, essentially a wingless rocket-powered jet fighter on wheels. In it, the intention is to reach a speed of 900mph with the sights set on going even faster, up to as high as 1,000mph.

Whether as conceptual, lead, system or detail designer, John Ackroyd's work on these multiform, high-profile schemes has called upon a broad understanding and application of a variety of fundamental physical laws, from aerodynamics to mechanics and the behaviour of materials, fluids and gases. Each, on his own admission, has confronted him with a steep learning curve but the mastering of these principles has singled him out.

Although not all of the many products designed by John Ackroyd were physically made on the Isle of Wight, all of his designs originated here. Moreover, the broad scope of his design engineering expertise is best revealed by highlighting the extensive range of these projects. The fact is that as a local man he has uniquely contributed much to the Isle of Wight's design and manufacturing heritage. As a personal tribute to John Ackroyd, *Aussie Invader* project leader Rosco McGlashen described him as an 'inspiration' and 'one of the cleverest design engineers, a truly remarkable man'.

Working for the Enfield Automotive Company at the Somerton Works, Cowes, John Ackroyd designed the first volume production electric car in the 1970s, the Enfield 8000 'city' car. Although the majority of the basic vehicles were assembled in a Greek plant after a wage dispute undermined the Island operation, ultimately 220 Enfield 8000 models were manufactured under a contract with the Electricity Council, the first leaving the production line in October 1973. The owner's manual was as simplified as the vehicle itself. (John Ackroyd)

DRIVER'S HANDBOOK
ENFIELD – 8000
ELECTRIC

ENFIELD AUTOMOTIVE LIMITED

SOMERTON WORKS, COWES, ISLE OF WIGHT, ENGLAND

Issue 2

It should be noted that the diminutive Enfield 8000 – it measured just 9.3ft in length with a 5.7ft wheelbase – was not inspired by environmental concerns but rather it was devised as a solution to the impact of the fuel oil crisis of the early 1970s which saw the price of a barrel of oil quadruple overnight with a commensurate increase in the cost of petrol at the pump. (John Ackroyd)

Exemplifying Island industrial innovation, the Enfield 8000 was photographed at Ryde, Isle of Wight, alongside a BHC SRN.6 hovercraft. A young John Ackroyd stands to the right, behind the car. (John Ackroyd)

The 8000 was not the only electric vehicle produced by Enfield Automotive. Among other designs was the Miner, a utility vehicle suited for underground operations where internal combustion engines presented a fire risk, and a leisure runabout called the Bikini. Both the Miner and the Bikini shared the chassis of the Enfield 8000 and were essentially stripped-down and modified versions of the passenger car with different bodywork. The Miner shown here was photographed in March 1976 at the Dinorwic Power Station in North Wales. Another was used in the Kiruna iron ore mines in Sweden's Arctic region. (John Ackroyd)

ARV Aviation, a company set up in Sandown by Richard Noble, was responsible for the ARV Super-2 light aircraft, named after the Air Recreational Vehicle application for which it was conceived. Designed by Bruce Giddings, with the talented John Ackroyd as a member of his team, it was constructed by novel techniques involving the bonding of 'Supral', a British-created superplastic aluminium alloy, using special adhesives to reduce rivet count and save weight. This resulted in a stronger, lighter airframe and a more affordable aeroplane for pilot training or private use. (Trevor Lyons)

John Ackroyd's greatest achievement was the design of the *Thrust 2* land speed record car. Incredibly for such a sophisticated, high-tech vehicle it was assembled in the old, rather run-down Ranalagh workshop at Wootton Creek. Driven by Richard Noble at the Black Rock Desert, Nevada, *Thrust 2* broke the world record on 4 October 1983 with a speed of 633.47mph. This view inside Ranalagh shows John Ackroyd, Ron Benton and Gordon Flux working on the incomplete car in 1980. (John Ackroyd)

Under the watchful eye of Richard Noble, *Thrust 2* is rolled out of the Ranalagh workshop in August 1981. Compare the condition of the low-tech workshop with that of the high-tech car, sleek and stylish despite lacking its final coat of paint and the sponsors' logos. (John Ackroyd)

The magnificent *Thrust 2*, resplendent in its gold livery, at Greenham Common in September 1981 with John Ackroyd sitting astride the engine casing. (John Ackroyd)

A month later, *Thrust 2* secured the record of fastest British car and driver when it achieved an average speed of 418mph over two runs on the Bonneville Salt Flats, Utah. (John Ackroyd)

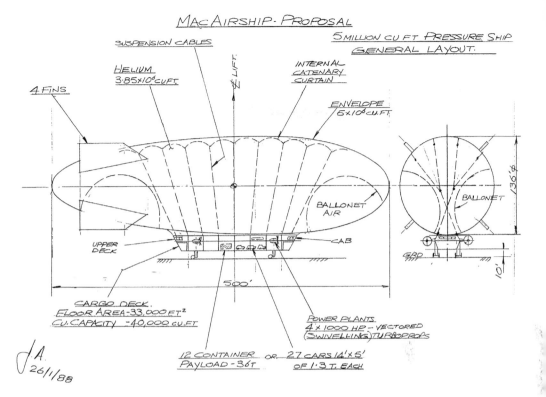

Above: An exceptional vehicle designed by John Ackroyd in 1988 was the helium-filled MAC Airbarge airship, intended for carrying freight and livestock from west to east Africa but unfortunately abandoned following the death of its commissioning customer. It would have measured 500ft in length with an envelope capacity of 5 million cu. ft. (John Ackroyd)

Opposite: The sponsors of the *Thrust 2* project derived enormous marketing benefit from their association with the record-breaking jet car and were keen to advertise the fact, in this case Trimite paint, the 'finest finish – fastest on Earth'. (Isle of Wight Heritage Services)

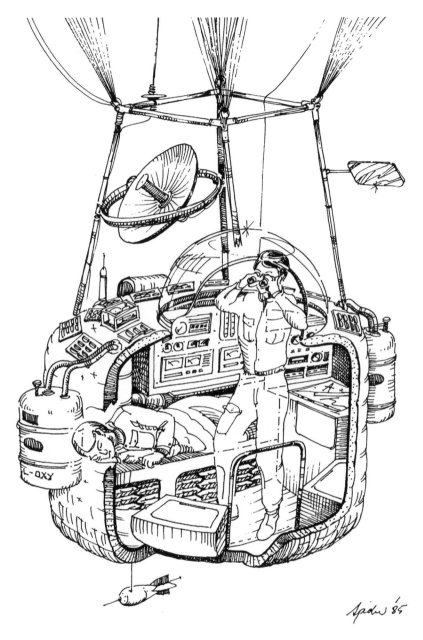

Apart from his work on land speed record cars, which continues with the *Aussie Invader 5R*, John Ackroyd has supported many ballooning record attempts, including those embarked upon by Sir Richard Branson, designing and supervising the construction of the pilot's capsules. This drawing, by Spider Anderson, the co-pilot, shows a cutaway view of the interior details of the capsule of *Endeavour*, a super-pressure balloon of 56,000 cu. ft volume intended to make the first global circumnavigation. (John Ackroyd collection)

Fabricated in two halves in Kevlar composite by Island Plastics at East Cowes, the upper section of the *Endeavour* capsule shell, with its reinforced aperture for the crew observation dome, is seen during production at East Cowes in 1984. (John Ackroyd)

Here the *Endeavour*'s capsule shell is shown bonded into a complete structure, with internal strengthening around the waistband. For such an endurance flight, some two to three weeks duration, the capsule would have been extremely cramped for its two occupants. Whereas a scaled-down prototype of the *Endeavour* balloon set records for its class flying over Australia in November 1984, the exhaustion of project funds led to the cancellation of *Endeavour*'s launch which had been due to take place in 1985 at the Pearce Airforce Base in Perth, Western Australia. (John Ackroyd)

John Ackroyd designed many other balloon capsules in his home office on the Isle of Wight, all of which performed successfully and helped to establish flight records. Photographed here post-flight are, on the left, the capsule of Sir Richard Branson's hot air balloon *Virgin Atlantic Flyer* and, on the right, the comparatively small capsule of the *Stratoquest* altitude record balloon. (John Ackroyd)

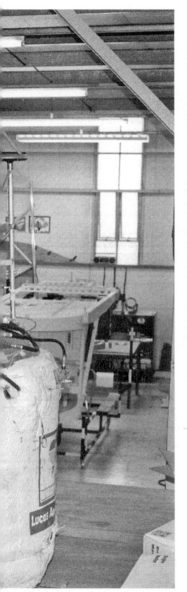

A night-time view of Sir Richard Branson and Per Lindstand's *Virgin Pacific Flyer* balloon prior to lift off, another ballooning record attempt for which John Ackroyd designed the all-important capsule. (John Ackroyd)

10

FLAT CAPS, BONNETS & OVERALLS
Working in Isle of Wight Industry

Undoubtedly, the conception of the range of amazing products shown in the preceding pages and the accomplishment of so many record achievements depended on the fertile and creative minds of those extraordinary designers and engineers, the stories of many of whom have been briefly related. The realisation of the end products would not have been possible, however, without the physical efforts of the many skilled, dedicated and loyal workers who have been employed in Island manufacturing businesses over the last two centuries. They were the ones who often bore the brunt of the bad times but who, despite the hardships, remained committed to these endeavours, continuing to be proudly involved even when laid off, and it is fitting to recognise their contribution to Isle of Wight industry.

The Island was, and to a large extent still is, a close-knit community. That would, of course, also apply to countless other industrialised areas around the country but, especially at West and East Cowes, there was an unusually strong camaraderie among the men and women who worked together at 'Sammy' Whites, Saunders-Roe and the other smaller boatyards and factories, perhaps accentuated because the operations were generally on a smaller scale than elsewhere and, living within the confines of an island with fewer manufacturing businesses, the workforce was less mobile. Not only did they work alongside each other in the yards, workshops and design offices but they socialised together in their home time, their children went to the same schools and they belonged to the same clubs, drank in the same pubs and so on. This communal intimacy extended into the

194

workplace creating something of a 'family' atmosphere and a strong sense of belonging to the project in hand, even if there was friendly rivalry between the different departments.

Nowhere was this more evident than on launch day at the shipyard or roll-out day at the aircraft works, when even those employees who were off duty would go into work in order to be part of the special occasion and the sense of achievement. And their wives and children were often there too.

As an example of the dedication exhibited by local working people, when Britten-Norman was in danger of failing to have its BN-2 Islander production prototype ready for the Farnborough Air Show of 1966, the entire 300-strong workforce forewent their annual holiday to ensure that it was completed. Rewarded for their commitment with a day's outing for the entire workforce, Desmond Norman said, 'The company has some of the finest craftsmen. They put their life and soul into ensuring perfection in every detail. Such an enthusiastic team could not be found anywhere else in the country.'

This 'family' culture was encouraged and supported by the directors and managers of each firm. It has left a lasting legacy too in the regular reunions of former apprentices and colleagues and retired staff groups. In the case of all the bigger Island employers, notably J. Samuel White, Saunders-Roe and Britten-Norman, all of them where finished craft construction took place, there was a caring attitude between and towards the employees which is not so evident these days, and at times of adversity, when cutbacks became unavoidable, considerable efforts were made to retain as many as possible through diversification.

Of course, this was not solely benevolent concern but influenced as much by a recognition of the value of these skilled workers who could be needed again in short order when things picked up; also an awareness of the investment that had been made in their training.

On more than one occasion, at times of downturn, local companies went to considerable lengths to try to preserve local employment until there were better prospects in the core business. After the First World War, Sam Saunders acquired the Gunville & Afton Brickworks and renamed it the Isle of Wight Pottery providing jobs for around seventy workers making high-quality ceramics – bowls, jugs, vases and other items – all now highly collectible. Later, following the loss of the SR.177 contract with its devastating consequences, there was an astonishing degree of diversity in the products and markets Saunders-Roe attempted to exploit. Rather than lay off all the men, some were even engaged making office furniture from surplus ply material accumulated at the Saro Laminated Wood Products plant at Whippingham.

Likewise, in the early 1960s, when the shipyard was under great threat, J. Samuel White enterprisingly embarked upon the development of numerous alternative product lines in a bid to stave off unemployment.

The Saunders-Roe Apprentices Training School at Osborne was second to none and, complete with its own residential accommodation, it attracted aspiring engineers, fitters and mechanics from all around the UK. The White's apprenticeship scheme was also gilt-edged, the company's indentures constituting a valuable testimony to the highest quality training that their holders had received, providing a passport to jobs with better prospects. Back in the nineteenth century, in the spirit of encouragement to his apprentices to advance their educations, John Samuel White set up a technical library from which all trainees were invited to borrow books.

The firms rewarded their employees in other ways too. Both White's and Saunders-Roe provided their staff with sports and social clubs and sporting competition under each company's respective identity was enthusiastically promoted and the results regularly published in staff magazines. There were sailing, cricket, football, tennis, rowing and boxing teams and annual sports days, partly involving light-hearted contests but also some rivalry with a serious competitive edge.

The employees also organised much of their own recreation and social activities taking full advantage of the facilities the companies had placed at their disposal. There was everything from dinner dances, launch parties, Christmas parties for staff children, annual balls and inter-departmental challenges in snooker, darts and bowls. At Saunders-Roe and, later, British Hovercraft Corporation, the annual themed Drawing Office (DO) Dinner was a popular and keenly supported entertainment event in which the men put on a stage show of sketches and songs, usually poking light-hearted fun at 'gaffs' that had been committed by colleagues and managers over the preceding twelve months, all taken in good part. Others organised fund-raising floats for the Island's carnival season or engaged in local charitable projects, while others arranged excursions and tours.

It was all these things which cemented the community bond of the groups of workers that served these Island companies, turning out ships, flying boats, hovercraft and aeroplanes, proudly committed to each end product to ensure its quality and success. Long Service Associations testify to the enduring service that many employees gave. Strange by today's transient standards, it was not uncommon, even up to the end of the twentieth century, for some men and women to have spent their entire working lives with the same employer.

Over the 200 or more years since the early 1800s, there have been countless people employed in Isle of Wight industry, the names of the majority now long forgotten other than by their families. The photographs that follow show some of these workers, both managers and staff, from relatively more recent times, from around the time of the First World War up to the present. Where they are known, the names of the persons depicted have been provided. This gallery of posed and informal photographs of those who were fortunate enough to have had their picture taken for posterity constitutes a token representation of the numerous men and women from each period who have been employed in local manufacturing. It is hoped they will be regarded as a tribute to all those many other workers who will remain unseen and anonymous but who, nevertheless, were as much an integral part of what was and still is 'Made on the Isle of Wight' as those who have been portrayed here.

J. Samuel White's Engine Works employees on 11 November 1918, in photographs taken to celebrate the armistice ending the First World War. In the larger group shot, apart from three men, they are all women. (Isle of Wight Heritage Services)

Boatbuilders from Saunders' Cornubia boatyard photographed in 1920. Many of Saunders' high-speed craft were constructed in the Cornubia yard. (Isle of Wight Heritage Services)

J. Samuel White Sports Club band from c. 1917. (Brian Greening collection)

Above: Folly workers from the Consuta production line in the early 1920s. Notice how female workers still dominate. (Philip Jewell collection)

Opposite top: The scene in one of Saunders' boatyards, believed to be part of the original Columbine Works, of the first Puma-type racing hull under construction. The design and production team pose with Samuel Saunders, standing at the front of the right-hand group. The man leaning on the hull on the opposite side is thought to be Fred Cooper. (Saunders-Roe, courtesy of Ray Wheeler)

Opposite bottom: A group from one of Saunders' boatyards engaged in lifeboat building for the RNLI. The youngest lad, seated at the front, can barely be 12 years old. (Saunders-Roe)

This is the team of men who built the *Miss England II*, photographed inside Saunders' small workshop in Union Street, now part of the modern Columbine Works site. (Author's collection)

A group of J. Samuel White's shipwrights photographed in 1934. Flat caps are very much the order of the day, even for the lads. Traditionally, foremen wore bowler hats as a mark of distinction. (Brian Greening collection)

A later view of Folly Works staff, dating from April 1932. At the centre in the front is Frederick William Baker, the general manager. Most of the women still wear a distinctive style of head covering. (Philip Jewell collection)

Female tracers in the Drawing Office at J. Samuel White in 1944. At the nearest two desks are Dolly Wise (left) and Elsie Nunn (right). (Brian Greening collection)

Above: The occasion of the Saunders-Roe Foremans & Staff Annual Meeting and Dinner at Queens Hall, Newport, on 8 March 1949. Among those who have been identified are Albert Gustar, Dennis Eckersley, Wilf Heber, Peter Robinson, Vic Denham, Harry Shepard, Wally Peters, Ray Garrett, Jack Priestley and Derek Hardy. At the top table are Mr Prefect and Captain Edward Clarke (managing director) with Sir James Milne, the managing director of J. Samuel White, who was presumably there as a guest. (John Farthing collection)

Opposite top: As the Second World War drew to a close, the staff of the Ship Fitting Section of J. Samuel White's MC Department were photographed together in June 1945. As in the First World War, with most men enlisted into the services, the continuation of production depended on female workers. (Richard de Kerbrech collection)

Opposite bottom: A wartime view of buses queuing in Folly Lane waiting to pick up Saro Laminated Wood Products staff at the end of a shift. With thousands of Island industrial workers, scenes such as this were typical well into the 1960s, not only at the Folly Works but also in Clarence Road, East Cowes, and Medina Road, West Cowes, providing transport for the shipyard and aircraft factory employees. (Brian Greening collection)

205

Above: More White's workers, these are Boatshop electricians photographed in 1956. They are: Len Blake, John Jackman, John Bennett, Fred Wright, Doug Toms, George Merton, Bob Cheek, Hubbard Hall, Dave McComb, Eric Griffin and Stan Tyler. Also present are the apprentices, Malcolm Stephens, Den Hamilton and Mick Hargreaves. (Brian Greening collection)

Opposite top: White's Engine Drawing Office team in the 1950s. Standing at the right-hand end of the middle row is Ron Trowell, a highly regarded local man who, though qualified as a chartered engineer, later became the company's official photographer. (Brian Greening collection)

Opposite bottom: White's Spar Shop workers, a photograph from the mid-1950s. From left to right in the back rows they are: Charlie Brake, Terry Whelan, Norman Docherty, Ken Hunnybun, unrecognised, Cyril Dallimore, Charlie Smith (a popular foreman), and Johnny Sleight; in the front: Roger Johnson, Terry Williams and Jacko Le Masurier. (Brian Greening collection)

Above: Some of the members of the Saunders-Roe team who were responsible for the world's first hovercraft, the SR.NI, standing outside the prominent Columbine hangar doors. From the left they are: Derek Hardy, Norman Davis, Phil Street, Vic Denham, Peter Lamb (chief test pilot), Chris Gear, Len Summers, Sir Christopher Cockerell (holding a model of the craft), unrecognised, Maurice Brennan (chief designer), Bill Crago, John Chaplin, unrecognised, Reg Davies, Doug Kemball, Peter Hayward, Ray Wheeler (deputy chief stress office and, later, chief designer) and Douglas Matthews. (Westland Aerospace, courtesy of Ray Wheeler)

Opposite top: Willing volunteers from the White's shipyard participate in a ship's lifeboat trial in the River Medina. Despite its serious purpose, as a break from work this activity was always a bit of fun. Across in East Cowes, BHC staff members also routinely took part in similar activities, testing the emergency evacuation procedures for new hovercraft. (Brian Greening collection)

Opposite bottom: Another photograph taken in White's shipyard, a group in the Turbine Shop in the 1960s. Among them are Mike Denham, John Williams and Edward 'Musket' Chapman. It is hard to imagine from the grubby overalls and working environment that some of these men were engaged in precision engineering. (Brian Greening collection)

Proudly photographed with the product of their labours, the Cushioncraft design and production team with the new CC-2 hovercraft in July 1961. John Britten and Desmond Norman are standing at the back at the right-hand end. The CC-2 was later much modified with the addition of a skirt and two external, boom-mounted engines. (Peter Gatrell)

Employees from EEL's Fluid Dynamics Labatories at a reunion in 2004. Photographed in the Sports & Social Club, the group of attendees include: Roger and Ann Ballard, Terry Bennet, Kay Brinton, Derek Cheal, Brian Clark, Bill Crago, Peter Dix, Dick Emerson, Tony Gazzard, Mary Grant, Brian Harris, Bob Harvey, Malcolm Keith, John Leonard, Joyce McGrath, George Marvin, Keith Newnham, Dave Perry, Cyril Pudan, Peter Seago, Martin Stevens, Steve Tutton, Stuart Welford, Percy Westwood and Dai Williams. (Steve Tutton collection)

A group of Britten-Norman office and production staff pose with Trislander G-BEPK in September 1984 prior to delivery of the aircraft to the Botswana Defence Force. (B-N Historians)

A team from J. Samuel White posing with a completed M8359 turbo gas expander. Among those pictured are: Keith Mitchell, Edward Chapman, Ken Deacon and Tony Westmore. (Ron Trowell)

Above: The British Hovercraft Corporation's Technical Publications and Airworthiness team from the early 1980s. From the left at the back: Peter Thorne, Michael Legg, Chris Bradley, Tony Devereaux, Yvonne Rampton, Ralph Compton, the author, Mike Bull, Graham Stephenson, Stuart Murray, John Edmunds, Ashley Hobbs, George Pimm, Graham Corbett, Greg (?) and Tim Skull; seated: Elaine Cesar, Sue Jenkins, Kay Martin, Sarah Wheeler and John Beaumont. (Author's collection)

Opposite top: The *Thrust 2* design, production and support team prior to the departure for Nevada in May 1983. Back row, left to right: John Watkins, Peter Hand, Gordon Flux, John Ackroyd (*Thrust 2* designer), Geoff Smee, David Tremayne, Glynne Bowsher, John Griffiths, David Brinn and John Norris; front row, left to right: Mick Chambers, Brian Ball, Ken Norris, Richard Noble (driver and team leader), Ron Benton, Mike Barrett, Eddie Elsom and Gordon Biles. (Chris Noble)

Opposite bottom: A gathering of Westland Aerospace Aviation Support and other staff in the 1990s. Among those in the photograph are, at the back from the left: Peter Edmunds, unidentified, Chris Colley, John Hickie, Trevor Burgess, Nick Taylor, Graham Corbett, Richard Dziedzicki, David Shorter, Peter Flynn, Tim Skull, Chris Franks, and Nikki Dominey; at the front from the left: David Parsonage, Brian Ash, Alan Bradley, Cara Spragg, Zoe Minter, Nikki Drummond, Maria Milford, Margaret Brooke, Sue Rose, Cathy Broadbridge, Caroline Crowie, Kay Brinton, and Pat Augustus. (Nick Henry)

Above: Prior to the takeover by GKN, the Westland Group Long Service Association regularly presented awards to staff from the Saunders-Roe, British Hovercraft Corporation, and Westland Aerospace eras (the sequence of names the company has held) in recognition of either twenty-five, forty or fifty years of service to the company. In this photograph, which dates from late 1994, production, technical and office employees receive their awards from Chris Gustar, managing director. Among them are: Steve White, John Crane, Stuart Hensall, Willie Merwood, Graham Corbett, Paul Cocker, Brian Bray, Derek Whitbread, John Fisher, Alan Scutt, Arthur Downer, Keith Smith, Dave Young, Ed Gouge, Justin Bell, Roy Middlebrook, Margaret Harlow, Val Taylor, Jim Young, David Soper and Mike Groves. (Author's collection)

Opposite top: The team responsible for the *Vestas Sailrocket 2* on launch day at the Venture Quays, formerly the Columbine Works, on 8 March 2011. In the rear are Kev Ellis, Glenn Thomas, Steve Pascoe, David Pratt and Shaun Kennedy. At the front are Rich Carter, Ian Pusey, Mick Hillman, Helena Darvelid, Paul Larsen and Ben Quemener. (Mark Lloyd, Lloyd Images)

Opposite bottom: Launch day! Always a major event in East and West Cowes going back for decades. Whether it was to see a new ship, flying boat or hovercraft, everyone turned out to see the product of the local workforce's labours. This scene, dating from 10 September 1934, shows the Royal Navy destroyer HMS *Fury* entering the River Medina. Every vantage point is occupied and the crowd includes many children and womenfolk. (Isle of Wight Heritage Services)

AFTERWORD

Returning to the question posed in the Introduction, 'What *has* been made on the Isle of Wight?' Apart from a great deal more, in summary there have been:

17 torpedo boats
106 destroyers & frigates
2,250+ fixed-wing aircraft
108 hovercraft
1,200+ lifeboats
42 space rockets (not all launched)

There have also been numerous other ships, hundreds of boats of all sizes, among them Cockleshell canoes and 'Goatleys', and thousands of aircraft components and assemblies in metal and composite, quantities that are constantly increasing with every working day. Besides many other record achievements, it has been the source of a world land speed record holder and six different world water speed record holders.

Much of this immense and innovative output still exists in one form or another as examples of the Isle of Wight's highly skilled design, engineering and manufacturing capabilities though, sadly, very little of it is now to be found locally. Instinctively, it is felt that the rightful place for such objects to be showcased for visitors, the local populace and even potential investors is somewhere on the Isle of Wight where they originated.

That said, among the larger surviving relics, the following can be found elsewhere:

The Polish super-destroyer ORP *Blyskawica* preserved in Gdynia, Poland

HMS *Cavalier*, the sole surviving Second World War Royal Navy destroyer at Chatham, Kent

A replica of the world water speed record holder *Miss England II* at the Science Museum, London

The rebuilt world water speed record holder *Bluebird* K3 in the Foulkes-Halbard collection at the Filching Manor Motor Museum, East Sussex

The steam gunboat *Grey Goose*, as the converted houseboat *Anserava*, at Hoo on the River Medway

The classic powerboat *Surfury*, in storage for the National Maritime Museum, at Wroughton, Wiltshire

Saunders-Roe's jet-powered flying boat, the SR.A1, on display at the Solent Sky Museum, Southampton. With it is the BN-1F Finibee and a replica Wight Quadruplane

The Saunders-Roe SR.53 mixed-power interceptor fighter at the RAF Museum, RAF Cosford, Wolverhampton

The *Black Arrow* space rocket R4 on display in the Science Museum, London

The original SR.N1 hovercraft, stored in the Science Museum Annexe at Wroughton, Wiltshire

The giant SR.N4 Mk III hovercraft in the Hovercraft Museum at Lee-on-Solent, Hampshire, along with numerous other Island-built air-cushion vehicles

The world land speed record holder *Thrust 2* in the Coventry Motor Museum

Although none of these preserved vehicles is located on the Isle of Wight, the Island is not entirely bereft of evidence of its creative past. Some of the birthplaces from which these and other products revealed in the preceding pages also still exist. Although much of the remnants of the J. Samuel White Engineering Works were destroyed in a fire in January 2016, the shipyard's distinctive hammerhead crane still stands out against the skyline at West Cowes. Across the River Medina in East Cowes, there is the giant Columbine works with its immense Union Jack doors which bear witness to the monster flying boats and hovercraft that once emerged through them. Alongside Bembridge airport, the Britten-Norman aircraft assembly works can be seen, both the original, by the Propellor Inn, and the new building, and the old rocket testing site at High Down, beyond Freshwater, remains facing the sea atop the cliffs, open to visitors.

There is also a valuable collection and display of items relating to the Island's maritime and aviation heritage at the Classic Boat Museum, now relocated to West Cowes. Its exhibits include the record-breaking transatlantic solo rowing boat *Britannia*, an example of the airborne lifeboat, a Mk VII Cockleshell canoe and Uffa Fox's Flying Fifteen yacht *Coweslip*, donated as a wedding present to Princess Elizabeth and Prince Philip in November 1947.

From 2017, fifty years after it made its maiden flight, an important addition to the foregoing artefacts will be the restored Islander aircraft G–AVCN, which will be on static display at Bembridge thanks to the efforts of the Britten–Norman Aircraft Preservation Society.

BIBLIOGRAPHY

Ackroyd, John, *Just for the Record: Thrust 2* (CHW Roles & Associates, 1984)

—*Jet Blast & the Hand of Fate* (Red Line Books, 2007)

Burdett, Sarah & Prior, Margaret, *East Cowes: A Town of Ships, Castles, Industry & Invention* (Dovecot Press, 2011)

Goodall, Michael H., *The Wight Aircraft* (Gentry Books, 1973)

Greening Brian, *Just Fifty Yards from the Floating Bridge* (Cowes Heritage, 2007)

Knowler, Henry, 'The Flying Boat Story', *SaRo Progress* (Winter 1954)

Levi, Renato, *Dhows to Deltas* (Nautical Publishing Co., 1971)

Marsh, George, *Britten Norman* (Tempus, 2000)

Rees, Quentin, *Cockleshell Canoes* (Amberley Publishing, 2010)

United States Navy, 'Report AD755-409: Air Cushion Vehicle Evaluation' (September 1971 to June 1972)

Wealthy, Bob, 'Saunders-Roe & the Princess Flying Boat' (talk given to Hamburg branch of the Royal Aeronautical Society, 3 June 2010)

—'Britten-Norman BN-2 Islander 50 Years On' (talk given to the Isle of Wight branch of the Royal Aeronautical Society, 16 June 2015)

Welford, Stuart E., 'Is it Right to Right?', *Naval Architect* (July 1974), pp. 93–7

Wheeler, Raymond, *Saunders-Roe* (Chalford, 1998)

—*A 20th Century Engineer* (Cross Publishing, 2010)

—*From River to Sea* (Cross Publishing, 1993)

Wheeler, Raymond & Tagg, Albert E., *From Sea to Air* (Crossprint, 1989)

Wheeler, Raymond & Chaplin, John B., *In the Beginning: the SR.N1 Hovercraft* (Cross Publishing, 2009)

Williams, David L., *White's of Cowes* (Silver Link, 1993)

—*Wings Over the Island* (Coach House Publishing, 1999)

Williams, David L. & Richard P. de Kerbrech, *J. Samuel White, Shipbuilder* (The History Press, 2012)

Wright, A.J. & Clancey, A.B., *Islander 96: Britten-Norman BN-2 Series Production History* (BN Historians, 1996)